Advanced Home Plumbing

Minnetonka, Minnesota, USA

Contents

Introduction 5

Planning Your Project. 7

Understanding the Water Cycle. 9
Mapping Your Plumbing System 14
Tools for Advanced Plumbing Projects 21
Plumbing Materials 22
Working with Copper Pipe 26
Working with Rigid Plastic Pipe 28
Working with Flexible Plastic Pipe 30
Working with Steel Pipe. 31
Working with Cast-iron Pipe 32
Understanding Plumbing Codes 34
Testing New Plumbing Pipes. 40

Formerly Cy DeCosse Incorporated
5900 Green Oak Drive
Minnetonka, Minnesota 55343
1-800-328-3895

Printed in U.S.A.

Books available in this series:
Everyday Home Repairs, Decorating With Paint & Wallcovering, Carpentry: Tools • Shelves • Walls • Doors, Kitchen Remodeling, Building Decks, Home Plumbing Projects & Repairs, Basic Wiring & Electrical Repairs, Workshop Tips & Techniques, Advanced Home Wiring, Carpentry: Remodeling, Landscape Design & Construction, Bathroom Remodeling, Built-in Projects for the Home, Kitchen & Bathroom Ideas, Refinishing & Finishing Wood, Exterior Home Repairs & Improvements, Home Masonry Repairs & Projects, Building Porches & Patios, Deck & Landscape Ideas, Flooring Projects & Techniques, Advanced Deck Building

Library of Congress
Cataloging-in-Publication Data

Advanced home plumbing.

p. cm.--(Black & Decker home improvement library)
Includes index.
ISBN 0-86573-750-9--
ISBN 0-86573-751-7 (pbk.)
1. Plumbing--Amateurs' manuals.
2. Dwellings--Remodeling--Amateurs' manuals.
I. Cowles Creative Publishing.
II. Series.
TH6124.A35 1997
696'.1--dc21 97-1176

COWLES
Creative Publishing, Inc.
Minnetonka, Minnesota, USA

President/COO: Nino Tarantino
Executive V.P./Editor-in-Chief: William B. Jones

Created by: The Editors of Cowles Creative Publishing, Inc., in cooperation with Black & Decker.

New Installation 42

Installing New Plumbing 44
Plumbing Bathrooms 48
Plumbing a Master Bathroom 50
Plumbing a Basement Bath 58
Plumbing a Half Bath 64
Plumbing a Kitchen 66
Installing Outdoor Plumbing 78

Replacing Old Plumbing 85

Replacing Old Plumbing 87
Evaluating Your Plumbing 88
Replacing Old Plumbing: A Step-by-step Overview 90
Planning Pipe Routes 92
Replacing a Main Waste-Vent Stack 96
Replacing Branch Drains & Vent Pipes 102
Replacing a Toilet Drain 106
Replacing Supply Pipes 108

Making Final Connections 111

Index 118

Executive Editor: Paul Currie
Senior Editor: Bryan Trandem
Associate Creative Director: Tim Himsel
Managing Editor: Kristen Olson
Project Manager: Lori Holmberg
Editors & Researchers: Jim Huntley, Joel Schmarje, Andrew Sweet
Editor & Technical Artist: Jon Simpson
Art Directors: John Hermansen, Gina Seeling
Senior Technical Production Editor: Gary Sandin
Technical Production Editors: Dan Cary, Greg Pluth
Copy Editor: Janice Cauley

Vice President of Photography & Production: Jim Bindas
Production Manager: Kim Gerber
Shop Supervisor: Phil Juntti
Set Builders: Troy Johnson, Rob Johnstone, John Nadeau
Production Staff: Curt Ellering, Greg Pluth, Brent Thomas, Kay Wethern
Studio Services Manager: Marcia Chambers
Photo Services Coordinator: Cheryl Neisen
Lead Photographer: Rex Irmen
Photographer's Assistants: Tom Heck, Mark Kolkman, Frederick Stroebel, Ingrid Worthman
Photographers: Kim Bailey, Greg Wallace

Contributing Photography: Hunter Industries; Kohler Co.; KraftMaid Cabinetry, Inc.; Metropolitan Council Environmental Services/St. Paul, MN

Printed on American paper by:
World Color
00 99 98 97 / 5 4 3 2 1

KOHLER

Introduction

Upgrading your home's plumbing system offers substantial rewards. Not only does new plumbing make your home more livable, it can also greatly increase its market value. Installing new plumbing is a perfect do-it-yourself project, because the work, though time-consuming, is not particularly difficult. Because plumbing contractors charge handsomely for their time, doing the work yourself can save you many hundreds or even thousands of dollars.

Advanced Home Plumbing is designed to meet the needs of a do-it-yourselfer with intermediate to advanced home-improvement skills and some experience working with plumbing materials. If you routinely perform your own minor plumbing repairs—replacing a faucet, fixing a toilet ballcock, patching an occasional leaky pipe—then this book is ideal for you. In *Advanced Home Plumbing*, you'll learn everything you need to know to install new plumbing lines, whether in a brand-new room addition or a converted space undergoing a major remodeling. You'll also learn techniques for replacing old steel and cast-iron pipes with new copper and plastic pipes.

Be aware that many plumbing projects require basic carpentry techniques. If you are unsure of your basic plumbing or carpentry skills, it is essential that you obtain good books on these subjects. Good reference sources include books such as *Home Plumbing Projects & Repairs* and *Carpentry: Remodeling*, from the Black & Decker® Home Improvement Library™.

Advanced Home Plumbing is divided into four easy-to-use sections.

The first section, Planning, provides important background information, and should be carefully read and fully understood before you attempt any plumbing project described later in the book. In this section, you'll learn the basic mechanics of the overall plumbing network, and how to inspect and map your existing plumbing system. You'll see the tools required for advanced plumbing projects and learn how to select and work with a variety of plastic and copper pipes and fittings. Finally, we'll discuss Plumbing Codes and explain how to work with building officials while planning your project.

The heart of the book lies in the next section, Installing New Plumbing. Four carefully chosen demonstration projects show how to install new plumbing in various situations—from a remodeled kitchen to a utility sink in a detached garage. No two projects are alike, so it is impossible for this book to cover every possible situation. When read as a group, however, these demonstrations show all the principles and techniques you're likely to need when completing your own plumbing project.

In the third section, Replacing Old Plumbing, we'll show you techniques for running new plumbing pipes when walls and floors are finished. You'll learn how to replace all plumbing pipes between the water meter and the individual fixtures. Some demolition and repair work is inevitable in any major plumbing project, but we'll explain the methods used by professional plumbers to save time and simplify this task.

The closing section, Making Final Connections, shows how to install all the fixtures and appliances in your home once the new drain pipes and water supply pipes have been installed. Dramatic cutaway models show every detail and leave nothing to chance.

Unique among how-to books, *Advanced Home Plumbing* takes the mystery out of understanding and installing a plumbing system. It is an essential addition to the library of any do-it-yourself homeowner.

NOTICE TO READERS

This book provides useful instructions, but we cannot anticipate all of your working conditions or the characteristics of your materials and tools. For safety, you should use caution, care, and good judgment when following the procedures described in this book. Consider your own skill level and the instructions and safety precautions associated with the various tools and materials shown. Neither the publisher nor Black & Decker® can assume responsibility for any damage to property or injury to persons as a result of misuse of the information provided.

The instructions in this book conform to "The Uniform Plumbing Code," "The National Electrical Code Reference Book," and "The Uniform Building Code" current at the time of its original publication. Consult your local Building Department for information on building permits, codes, and other laws as they apply to your project.

Planning Your Project

In a basic plumbing cycle, which is still used in many rural homes, a well pumps fresh water from the water table. A system of in-house plumbing pipes then distributes the water to various fixtures throughout the house. After use, the waste water travels through a system of drain pipes to a septic system, which separates solid wastes and returns the water to the soil. In metropolitan areas, the methods for supplying fresh water and handling waste water are more complex (see pages 10 to 11).

Understanding the Water Cycle

Each day in the United States, roughly 500 billion gallons of water are withdrawn from surface and underground water supply sources for use in homes and businesses. In private homes, water is used for drinking and food preparation, for cleaning and washing, and for irrigating landscapes and gardens. Finally, we depend on water to carry septic wastes safely away from our homes.

Just as fresh water is essential to basic human existence, a well-designed plumbing system is essential to a healthy life and productive society. The science of transporting water to and from a home through a system of pipes originated some 5,000 years ago in ancient Sumer and is still evolving today.

Most homeowners are familiar with the in-home portion of the water cycle—the plumbing system with its network of faucets, toilets, and other plumbing fixtures. Many homeowners can make minor repairs to the fixtures and pipes. However, few homeowners fully understand the beginning and ending stages of this sequence—called the hydrologic cycle. Knowing where the fresh water supply originates can help you make decisions on how to manage the water entering your home; and understanding how waste water is recycled may change your decisions on what materials to flush down the drain.

Distributing fresh water. The fresh water that enters your home originates in one of two sources: surface water or groundwater. Surface water supplies include lakes, streams, rivers, and artificial storage reservoirs. Groundwater comes from natural underground caverns or from aquifers—porous, water-saturated layers of gravel, sand, and silt. This water is either pumped from the ground through a well, or rises to the surface through natural springs.

In about 25% of homes, mostly rural residences, a well delivers fresh water directly from groundwater sources with no chemical treatment. This is done because groundwater is generally regarded as pure enough to drink. In recent years, however, it has been found that some groundwater supplies have unhealthy levels of natural substances, such as nitrates, as well as toxic pollutants and dangerous microorganisms. If your water supply comes from a well, it is a good idea to have a sample tested by a laboratory. If a test finds that your water supply has harmful contaminants, consider installing an in-home purification or filtration system.

In addition, groundwater pumped from a well often has a high mineral content. This "hard" water may prematurely age your plumbing pipes as the minerals build up on the inside surfaces of the pipes. Installing a water softener helps reduce mineral levels and can extend the life of your plumbing system.

In metropolitan areas and many smaller communities, fresh water distribution is a function of a public utility. The water pumped from the ground or taken from surface sources is first cleansed at a central treatment plant before it is distributed to homes. Public utilities generally do an adequate job of purifying the water supply, but in large industrial areas, especially those that derive their water supply from surface water, it is not uncommon for very low levels of potentially harmful substances to be present. Your local division of the Environment Protection Agency should have information on the water quality in your area.

Recycling waste water. About 78% of all water that enters a home or business eventually finds its way back to a groundwater or surface water source. Because this waste water eventually becomes part of the fresh water supply again, it must be purified before it is released into the environment.

In rural areas, this purification is accomplished by simple home septic systems that separate solid wastes and transport the water back into the soil. As this water makes its slow journey back to the water table, it is filtered pure by many layers of rock and soil.

In most cities and towns, waste water purification is accomplished by a system of sewer utility pipes that carry raw sewage to a central treatment facility. After solid wastes are removed, the water is purified and released.

Keep in mind that individual septic systems and urban sewage treatment plants are designed to recycle waste water and organic solids only. Flushing any synthetic materials or chemicals down your drain can jeopardize the system's function—transforming waste liquid into pure fresh water for the next user.

Man-made reservoirs created by constructing dams across river valleys are common in arid western regions. By capturing water from melting snows in the mountains, communities can store enough water to supply their needs through the dry summer months. In other regions, rivers and lakes provide ongoing sources of fresh water. Many man-made reservoirs are hydroelectric projects used to generate electricity, as well as to store fresh water.

Water towers and enclosed tanks store the water pumped from treatment plants. They are usually positioned on high ground, so gravity can produce the downward pressure necessary to force water through the water mains. Every 2.3 feet of height produces 1 pound of water pressure.

Distributing Fresh Water

Although the specific methods for collecting, treating, and distributing fresh water vary from community to community, the process is generally the same no matter where you live. Water is pumped from an underground source or is channeled from a lake, river, or reservoir into a large controlled storage facility.

From the storage reservoir, the water is pumped to a water treatment plant, where it is cleansed and purified. Water treatment plants use a variety of physical and chemical procedures. In a typical plant, water is strained and filtered to remove solids, is aerated with sprayers to remove dissolved gases, and is then disinfected with chlorine to kill organisms. In most communities, fluoride is added to the water supply to help reduce tooth decay.

Finally, the treated fresh water is pumped to enclosed storage towers or tanks, where a network of distribution pipes carries the water to homes and businesses in the community.

Recycling Waste Water

The systems used to purify waste water after it leaves the home range from simple home septic tanks to enormous sewage treatment plants that handle millions of gallons of raw sewage each day. But no matter what system is used, the process of cleansing waste water is simple in principal.

All waste treatment systems use a combination of physical, biological, and chemical processes to remove contaminants from water. However, the success of any waste water treatment system can be compromised by improper use. Bleaches, fertilizers, and phosphorus detergents flushed into the waste system may interfere with the natural consumption of solid wastes by microorganisms. And pesticides and other chemicals poured down sink drains by homeowners may eventually find their way to the fresh water supply.

Lagoon systems are growing in popularity, especially for small communities. Originally, this design was used simply to hold untreated waste water until it evaporated, but modern lagoons are state-of-the-art treatment facilities in which aquatic plants are used to break down solid wastes.The plant material is periodically harvested and dried to create fertilizer, and the water is used to irrigate farming lands or urban landscapes.

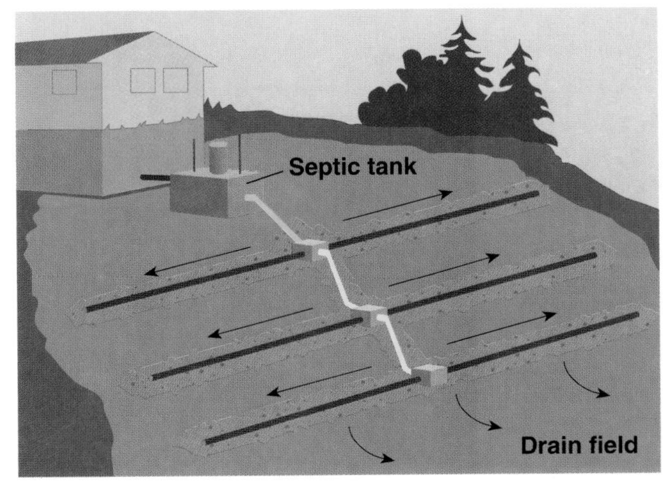

Private septic system consists of an underground tank and a system of pipes fanning out from the tank. When sewage reaches the tank, the solid wastes settle to the bottom, where they are consumed by microorganisms. As the tank fills, the water flows out of the tank through porous drain pipes that distribute the water into the soil. The water is filtered clean as it drains down through thick layers of soil and rock on its return to the water table. Used correctly, a septic system requires only that residual solid wastes be pumped out every few years.

Sewer treatment plant is the popular urban solution for treating billions of gallons of waste water that drain from our homes and businesses. Sewage, a combination of water and solid wastes, is first directed to settling tanks, where the solid wastes precipitate to the bottom. This is followed by a biological treatment, in which microorganisms digest remaining organic material in the water. The water is then filtered, disinfected with chlorine, and discharged to irrigation canals, streams, or lakes. The residual solid waste is often processed as agricultural fertilizer.

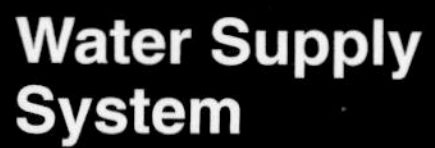
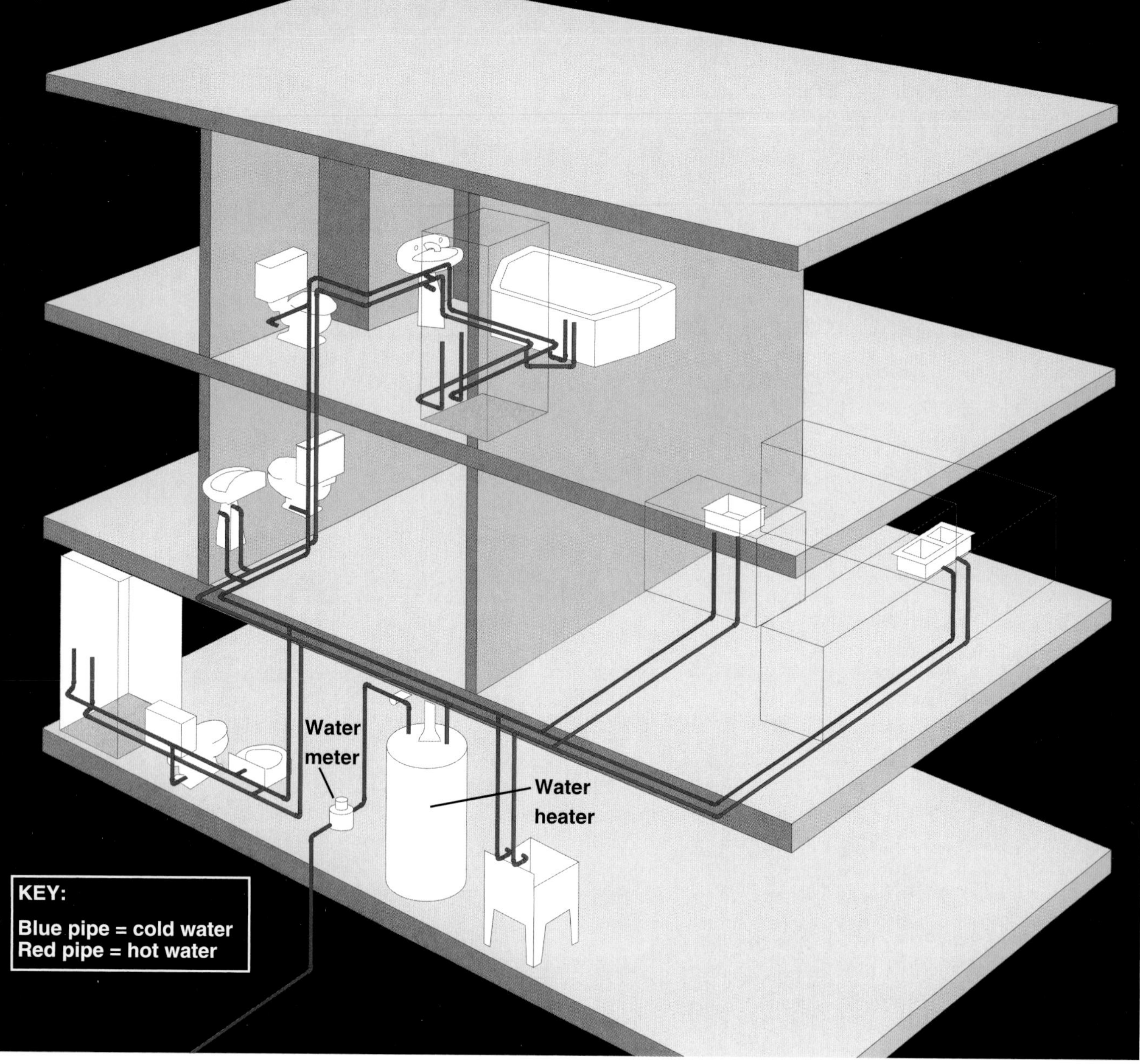

Using Water in Your Home

Behind the walls, under the floors, and beneath the landscape, a typical home hides a seemingly chaotic system of pipes, drains, hoses, tubes, sprinklers, valves, elbows, traps, bends, valves, faucets, fixtures, and spigots—enough to nearly circle a football field, if all the pipes and fittings were laid end to end.

But no matter how bewildering the plumbing network appears, its purpose is simple: to carry fresh water to points of use and to transport waste water out of your home. All home plumbing systems consist of two separate systems: the water supply system (above) and the drain-waste-vent (DWV) system (opposite page).

Water supply pipes are relatively small—1" or less in diameter. They form a tightly sealed, pressurized system controlled by valves and faucets. Older homes may have galvanized steel supply pipes, while newer homes (or renovated older homes) have copper or plastic supply pipes.

Upon arriving in your home, the 1" main water supply pipe passes through the water meter and begins to split into ¾" or ½" branch lines almost immediately. One branch passes into a water heater, which will supply hot water for faucets and other fixtures in the home. Hot and cold water branch pipes run parallel to one another on their way to kitchens, bathrooms, and utility sinks

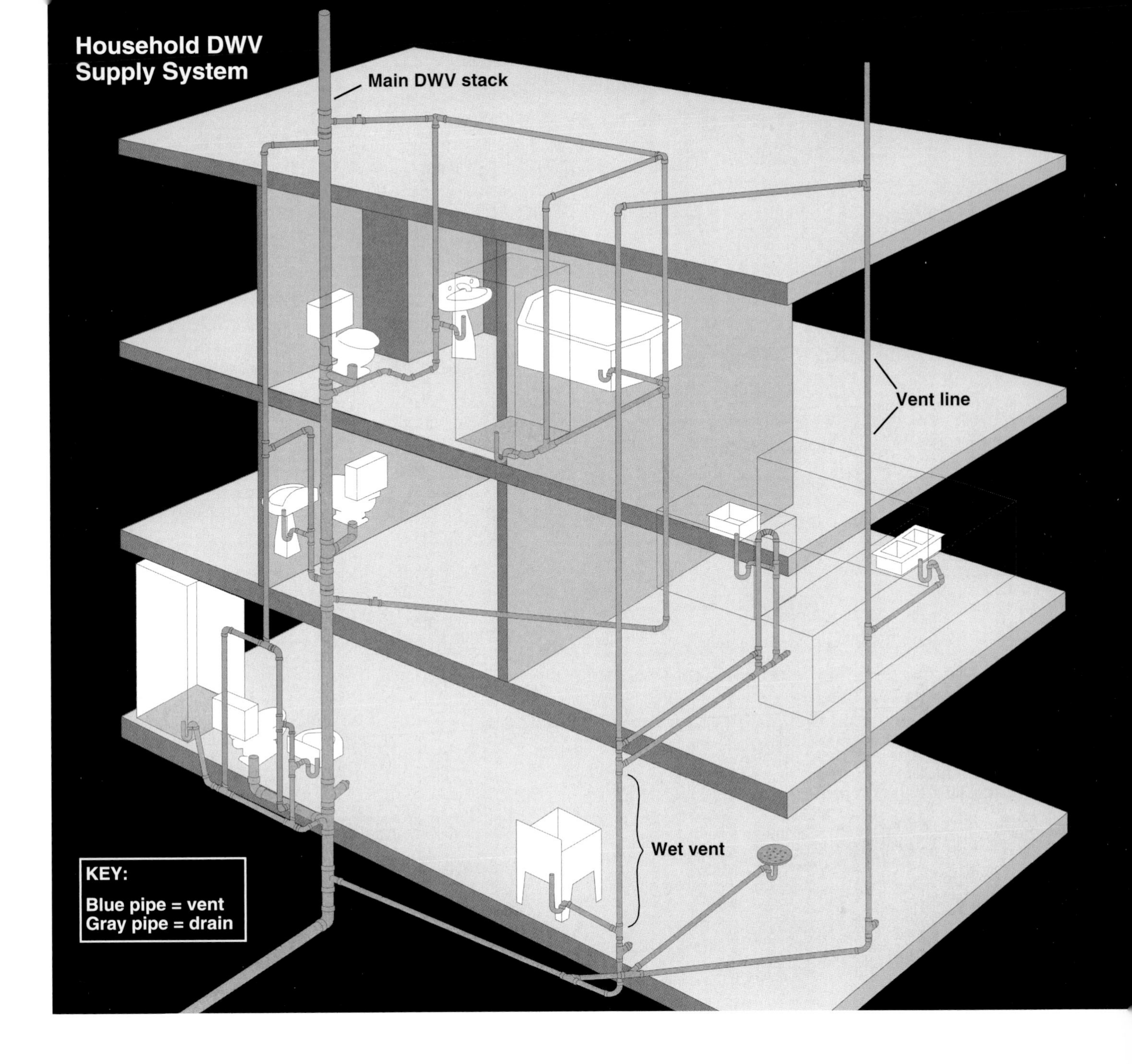

throughout the home. Other branch lines, supplying only cold water, extend through the foundation to outdoor hose bibs and detached outbuildings.

DWV pipes, unlike the sealed supply system, form an open system ventilated to the outdoors. At 1¼" to 4" in diameter, DWV pipes are larger than the water supply lines. In older homes, DWV pipes are typically cast iron or galvanized steel, while in newer houses they are usually plastic.

To prevent noxious gases from rising into your home from the sewer, each fixture in your DWV system is serviced by a looped section of pipe, called a *drain trap*. The trap holds standing water, effectively sealing the sewer system off from the interior of your home. If the system were sealed, flowing waste water would create a partial vacuum that could suck the standing water out of drain traps. To keep the system open and prevent this problem, each fixture drain is connected to a nearby vent line, which is linked to a waste-vent stack extending through the roof. By equalizing pressure in the system, the vent pipes ensure the proper function and safety of your drain system.

Mapping Your Plumbing System

Find supply lines inside walls or beneath floors by listening for running water with a stethoscope or a drinking glass while a helper runs water from the fixture. Use fixture locations on the floors above and below to find the general location of pipes.

Mapping your home's plumbing system is a good way to familiarize yourself with the plumbing layout and can help you when planning plumbing renovation projects. With a good map, you can envision the best spots for new fixtures and plan new pipe routes more efficiently. Maps also help in emergencies, when you need to locate burst or leaking pipes quickly.

Draw a plumbing map for each floor on tracing paper, so you can overlay floors and still read the information below. Make your drawings to scale and have all plumbing fixtures marked. Fixture templates and tracing paper are available at drafting supply stores.

Tips for Mapping Your Plumbing System

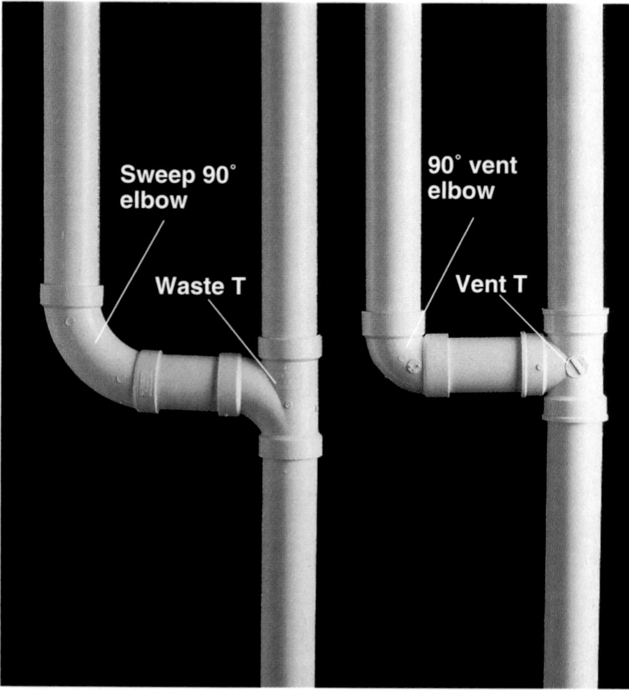

Identify drains and vents by the shape of their fittings. Drain pipes (left) require gradual changes in direction, requiring the use of Y-fittings, waste T-fittings, and sweep 90° elbows. Vents (right) can use fittings with abrupt changes in direction, such as vent T-fittings and vent elbows.

Pinpoint the location of the main stack when all interior walls are finished by jiggling a hand auger down the roof vent while a helper listens to walls from inside the house. Always exercise caution when working on a roof.

Use floor plans of your house to create your plumbing map. Convert the general outlines for each story to tracing paper. The walls can be drawn larger than scale to fit all the plumbing symbols you will map, but keep overall room dimensions and plumbing fixtures to scale. Be sure to make diagrams for basements and attic spaces as well.

Locate and map all valves throughout the supply lines. This will allow you to shut off only the necessary branches when making repairs, while maintaining service to the rest of the house. Use the correct symbols (right) to identify different valve types (page 23).

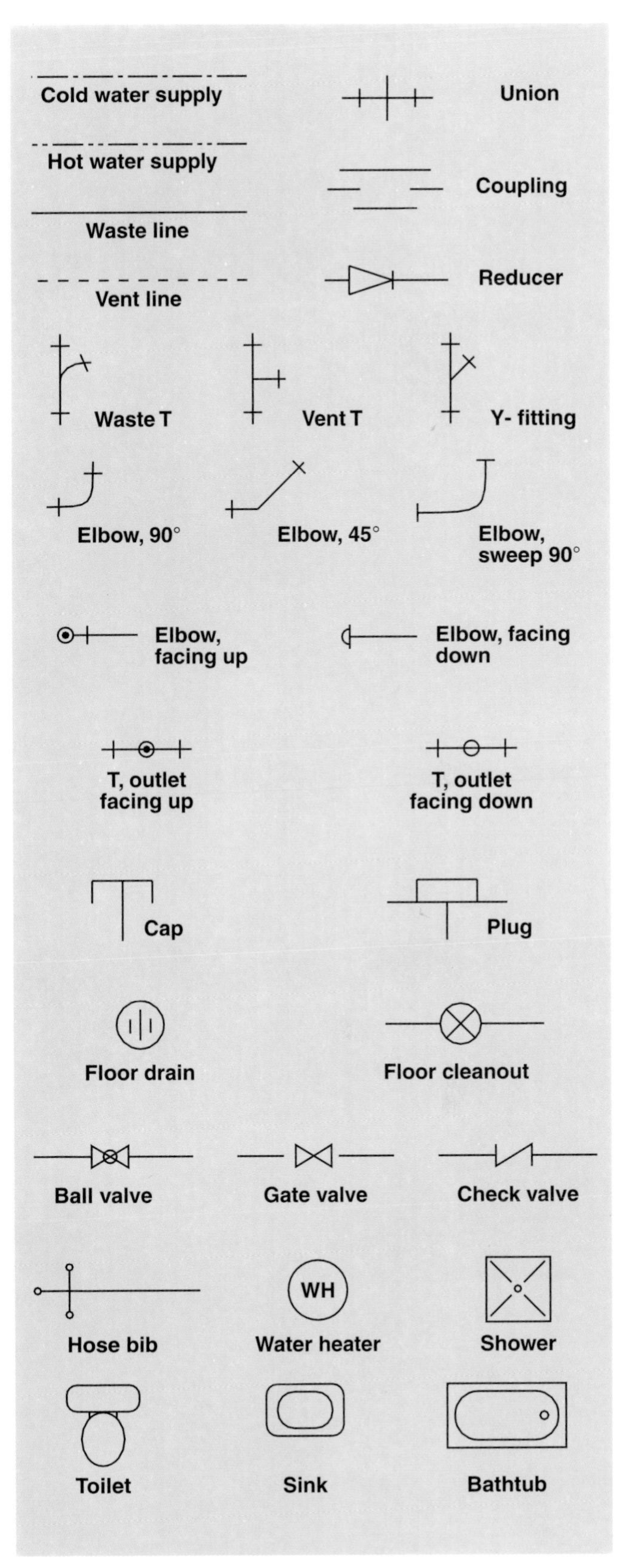

Use standard plumbing symbols on your map to identify the components of your plumbing system. These symbols will help you and your building inspector follow connections and transitions more easily.

How to Map Water Supply Pipes

1 Locate the water meter, usually found along a basement wall. The meter is the first main fitting in the supply line. Mark its location on your basement diagram.

2 Follow the cold water distribution pipe past the main shutoff valve to the water heater, generally the first appliance to receive water. Map the valve and water heater locations on the basement diagram.

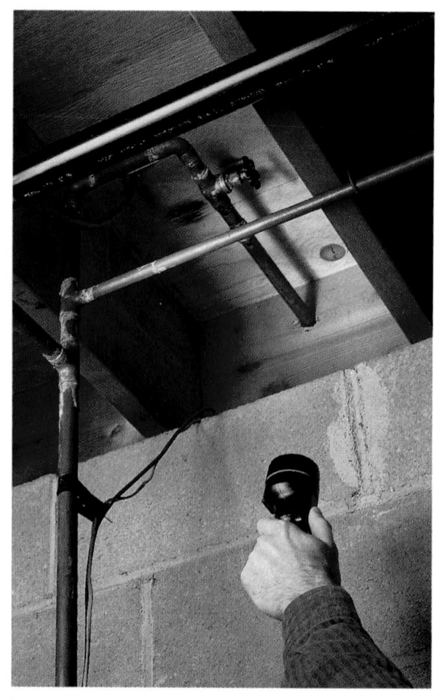

3 Locate cold water branch pipes leading to sillcocks, which supply water to hose bibs outside the house. Indicate these branches on the basement map.

4 Return to the water heater and map the location of hot and cold supply lines running to basement utility fixtures, such as a washing machine and utility sink.

5 Map the routes to any remaining basement plumbing fixtures on your basement diagram. Pipe runs that serve both basement and first-floor plumbing should be marked on both the basement and first-story maps.

6 Find where vertical supply pipe risers extend up into the floors above. Supply lines generally rise straight to the first-floor fixtures. If there is doubt, measure from the nearest outside wall to the supply line riser, and do the same at the respective first-floor fixture. If measurements are not the same, there is a hidden offset in the pipe route.

7 Map the supply routes to all first-story fixtures by laying the first-floor diagram over the basement map and transferring the locations of vertical supply risers. Indicate any jogs in the supply lines occurring between floors.

8 Overlay the second-story diagram over the first-floor map, and mark the location of supply pipes—generally they will extend directly up from fixtures below. If first-story and second-story fixtures are not closely aligned, the supply pipes follow an offset route in wall or floor cavities. By overlaying the maps, you can see the relation and distance between fixtures and accurately estimate pipe routes. If no obvious path exists for supply pipes, try locating the pipes by listening in likely areas with a stethoscope (page 14).

How to Map DWV Pipes

1 From the basement, locate the main waste-vent stack and any fixtures that drain directly into it, such as a basement toilet.

2 Determine the path of the main drain under the basement slab by following the main stack to the cleanout hub on the basement floor. The cleanout is usually located near a basement wall facing the street.

3 Note any auxiliary waste-vent stacks that enter the basement floor. These are typically 2"-diameter pipes, compared to 3"- or 4"-diameter main waste-vent stacks. Auxiliary waste-vent stacks are often located near basement utility sinks or below a kitchen located far from the main stack.

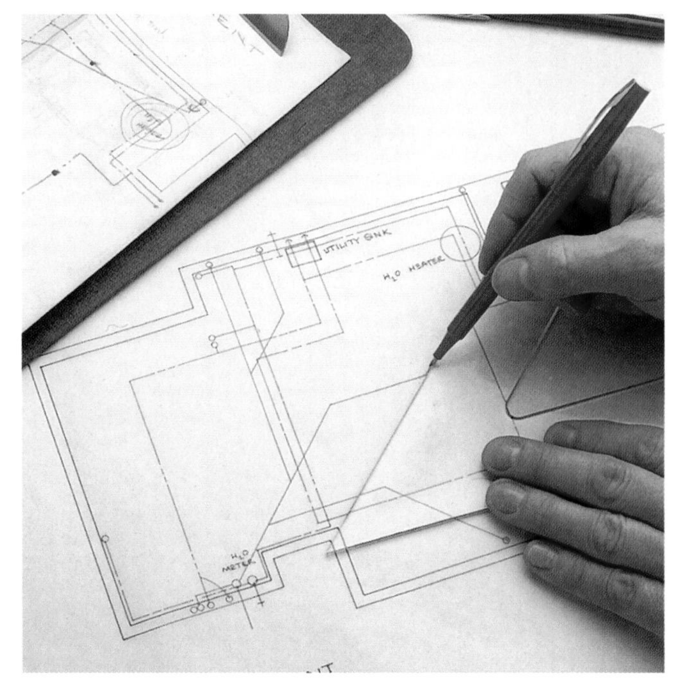

4 On your basement diagram, map the location of the main waste-vent stack, the cleanout hub, and the horizontal main drain pipe. Also note the location of auxiliary stacks, and estimate the path of the horizontal drain pipe connecting the auxiliary stacks to the main drain.

5 From the basement, note the location of horizontal drain pipes running overhead, and the points where vertical drain pipes extend up into the floors above. Overlay your first-story diagram onto the basement map, then transfer the location of the vertical waste-vent stacks. Mark the location of all horizontal drain pipes running below the floor.

6 Overlay the second-story diagram over the basement map, transfer the location of the vertical waste-vent stacks, and mark the location of any horizontal drain pipes running beneath the floor. Since the floor spaces between the first and second story are usually finished, you may need to estimate their location. These horizontal drain pipes will usually drain into the nearest waste-vent stack.

7 Finally, map the location of vent pipes as accurately as you can. If possible, look in your attic to determine where vent pipes emerge from the story below. Indicate whether the individual vent pipes connect to a waste-vent stack or extend through the roofline.

A
Pipe wrenches
Adjustable wrenches
Channel-type pliers
Ratchet wrench
Hammer
Plastic tubing cutter
Wire brush
Propane torch
Flux brush
Utility knife
Copper tubing cutter
BERNZOMATIC
PROPANE
FUEL CYLINDER
Net Wt. 14.1 oz. (400 gm.)
B

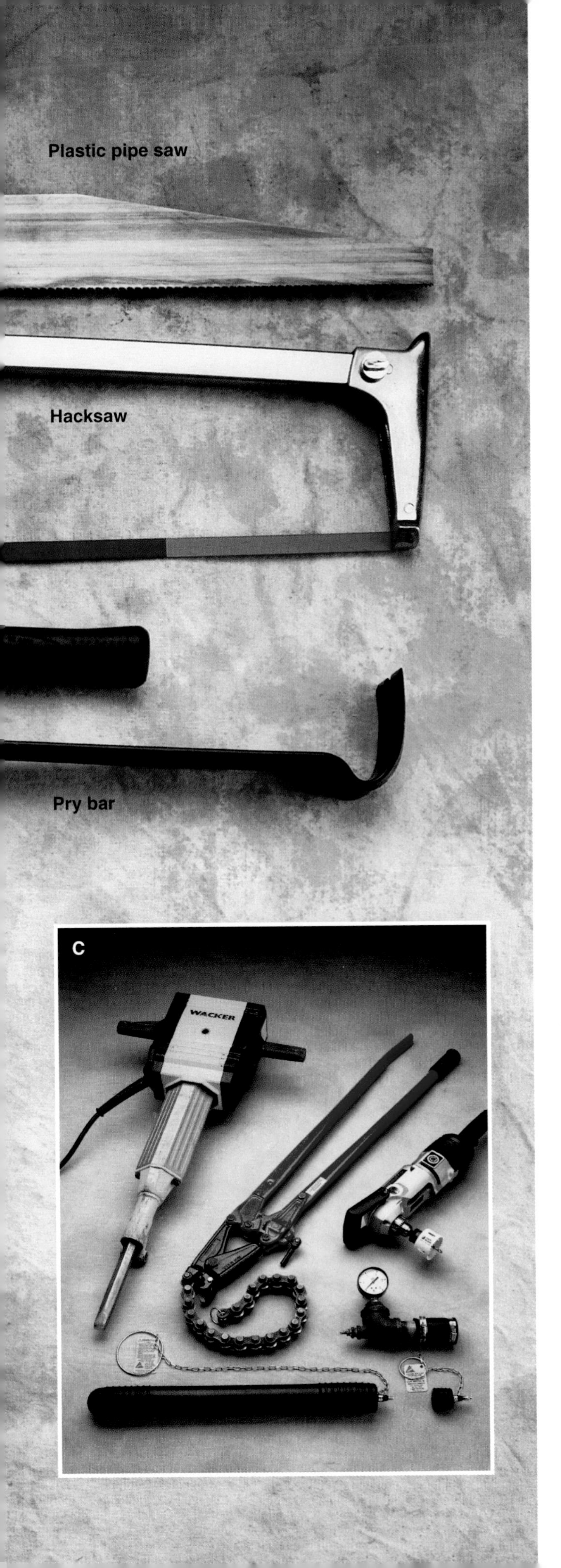

Tools for Advanced Plumbing Projects

The tools shown in these photos will be used in the advanced plumbing projects demonstrated in this book. We recommend that all do-it-yourself plumbers own the basic hand and power tools shown here (photos A, B), because you will use them extensively. If you are purchasing new tools, always invest in quality.

Occasionally, specialty tools will be needed (photo C). These can be leased inexpensively at most rental centers.

In addition, some plumbing projects require demolition or construction work. If your project falls into this category, you should also own a full complement of basic carpentry tools, including a tape measure, level, saw, plumb bob, masonry chisel, and maul.

Tools used in plumbing will very likely be exposed to water. Prevent rust by wiping your tools dry and applying a light coat of household oil after using them. Lubricants available in spray cans are convenient for this purpose.

Power tools you will find useful for completing advanced plumbing projects are shown in photo B. They include (from top left): power miter box, circular saw, jig saw, drill, and reciprocating saw.

Rental tools can make plumbing projects much easier to complete. Tools available for lease at rental centers are shown in photo C. They include (from top left): electric jackhammer for breaking concrete, cast iron snap cutter, right-angle drill with hole saw, and test kit for pressure-testing DWV pipes.

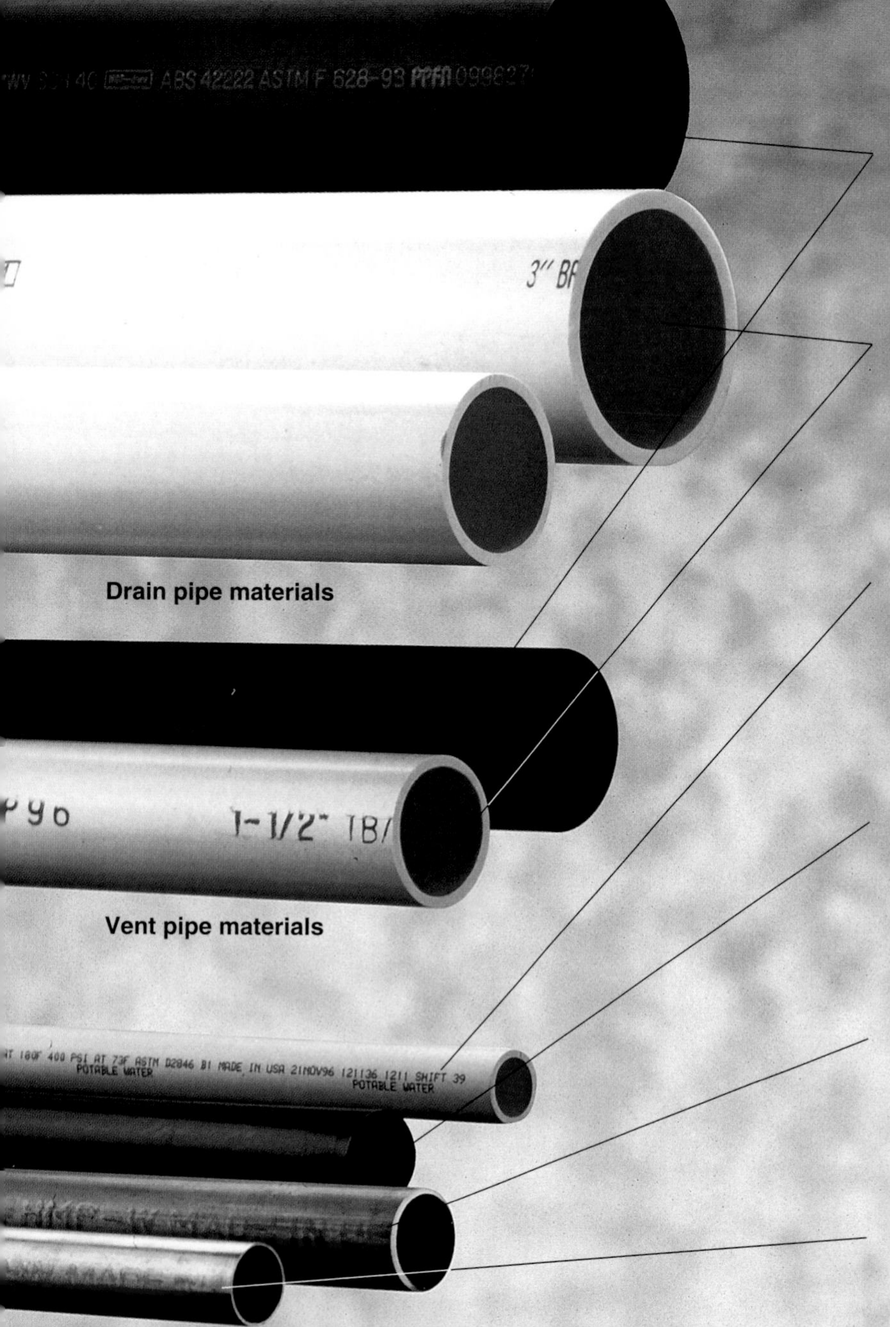

Drain pipe materials

Vent pipe materials

Supply pipe materials

Common Pipe Materials

Schedule 40 ABS (acrylonitrile butadiene styrene) plastic is a rigid black or dark gray pipe used for drain and vent lines. It is commonly available in 10-ft. and 20-ft. lengths, in diameters of 1½", 2", 3", and 4".

Schedule 40 PVC (polyvinyl chloride) plastic is a white or cream-colored rigid pipe most commonly used for drain and vent lines. PVC is sold in 10-ft. and 20-ft. lengths, in diameters of 1¼", 1½", 2", 3", and 4".

CPVC (chlorinated polyvinyl chloride) plastic is a cream-colored pipe that in some areas is approved for use in hot and cold water supply lines. Where allowed, CPVC should be rated for 150 psi and 210°F when used in water supply lines. It is commonly sold in 10- ft. and 20-ft. lengths, in diameters of ½", ¾", and 1".

PE (polyethylene) plastic is a black or bluish flexible pipe that is often used for main water service lines from the street to the home, and for outdoor cold water pipes, such as those used for landscape watering systems. PE pipe is sold in coils of 50 ft. or more, in ½", ¾", and 1" diameters.

Type-L copper is a thick-walled pipe used primarily for underground water supply lines. Type-L is available both in rigid and semi-flexible form. Rigid copper is sold in 10-ft. and 20-ft. lengths; flexible copper in 60-ft. coils. Type-L copper is available in ½", ¾" and 1" diameters.

Type-M copper is the standard for indoor water supply lines. It has thinner walls than type-L copper, making in less expensive and easier to cut. It is available in 10-ft. and 20-ft. lengths, in diameters of ½", ¾" and 1".

Plumbing Materials

The materials used in home plumbing systems are closely regulated by Building Codes. The materials shown here and on the following pages are approved for use by the National Uniform Plumbing Code at the time this book was published. However it is possible that your local code has other requirements, so always check with your local building officials. Approved materials are stamped with one or more product standard codes. Look for these stamps when buying your pipes and fittings.

- Working with Copper Pipe (pages 26 to 27)
- Working with Rigid Plastic Pipe (pages 28 to 29)
- Working with Flexible Plastic Pipe (page 30)
- Working with Steel Pipe (page 31)
- Working with Cast-iron Pipe (pages 32 to 33)

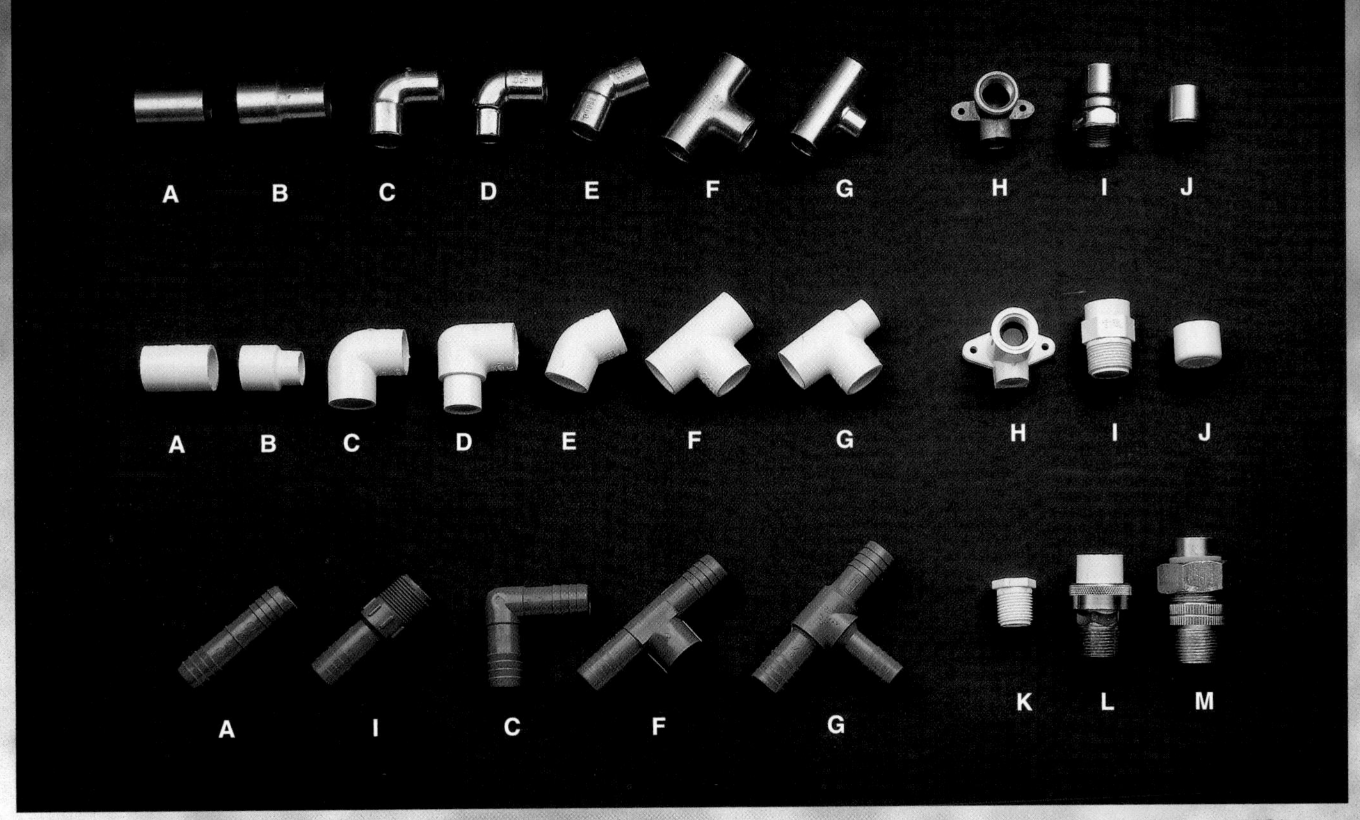

Water supply fittings are available in copper (top), CPVC plastic (center), and PVC plastic (bottom). PVC water supply fittings are gray with barbed sleeves, and are used only with cold water PE pipe. Fittings for each material are available in many shapes, including: unions (A), reducers (B), 90° elbows (C), reducing elbows (D), 45° elbows (E), T-fittings (F), reducing T-fittings (G), drop ear elbows (H), threaded adapters (I), caps (J), plug (K), CPVC to copper transition (L), and copper to steel transition (M).

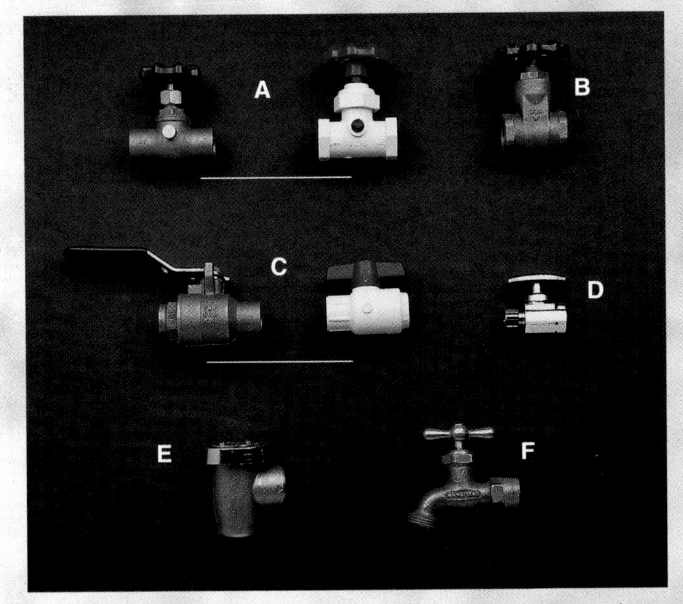

Water supply valves are available in bronze or plastic and in a variety of styles, including: drain-and-waste valves (A), gate valve (B), full-bore ball valves (C), fixture shutoff valve (D), vacuum breaker (E), and hose bibb (F).

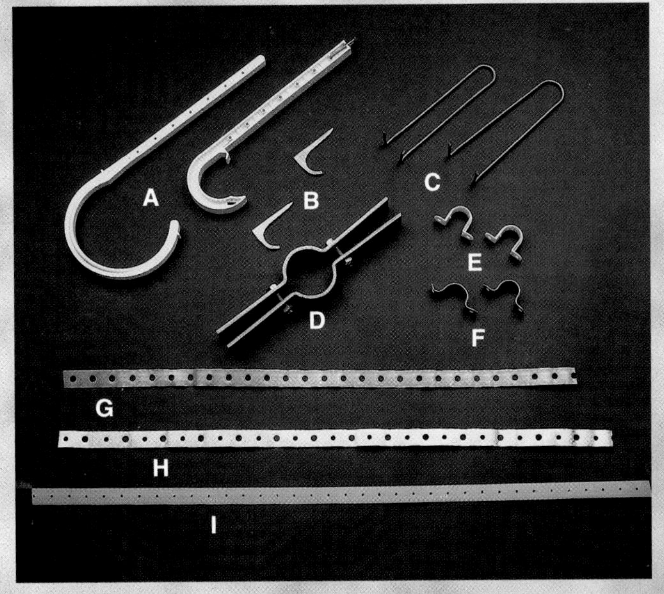

Support materials for pipes include: plastic pipe hangers (A), copper J-hooks (B), copper wire hangers (C), riser clamp (D), copper pipe straps (E), plastic pipe straps (F), flexible copper, steel, and plastic pipe strapping (G, H, I). Do not mix metal types when supporting metal pipes: use copper support materials for copper pipe, steel for steel and cast-iron pipes.

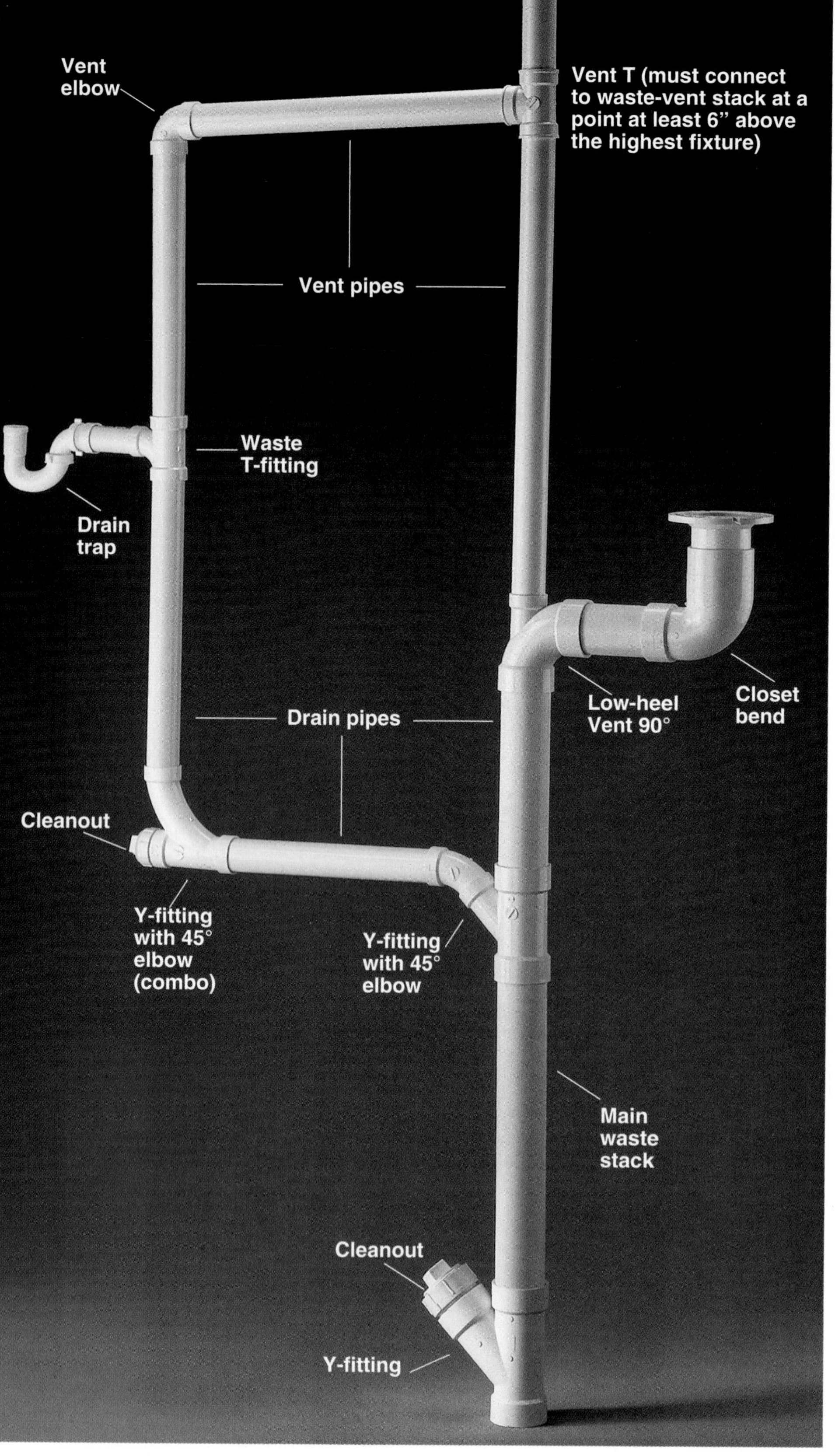

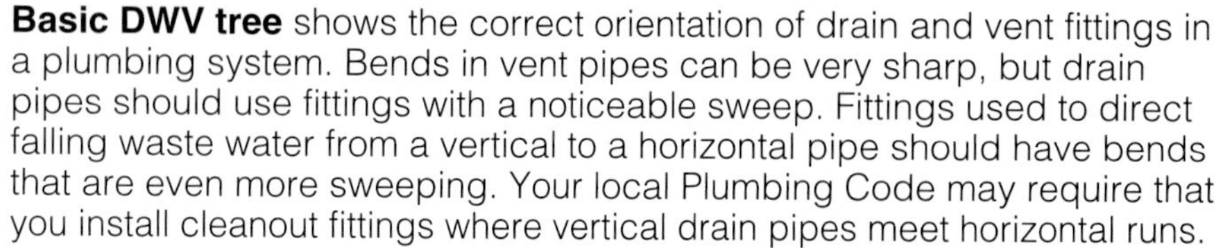

Basic DWV tree shows the correct orientation of drain and vent fittings in a plumbing system. Bends in vent pipes can be very sharp, but drain pipes should use fittings with a noticeable sweep. Fittings used to direct falling waste water from a vertical to a horizontal pipe should have bends that are even more sweeping. Your local Plumbing Code may require that you install cleanout fittings where vertical drain pipes meet horizontal runs.

DWV Fittings

Use the photos on these pages to identify the DWV fittings specified in the project how-to directions found later in this book. Each fitting shown is available in a variety of sizes to match your needs. Always use fittings made from the same material as your DWV pipes.

DWV fittings come in a variety of shapes to serve different functions within the plumbing system.

Vents: In general, the fittings used to connect vent pipes have very sharp bends with no sweep. Vent fittings include the vent T and vent 90° elbow. Standard drain pipe fittings can also be used to join vent pipes.

Horizontal-to-vertical drains: To change directions in a drain pipe from the horizontal to the vertical, use fittings with a noticeable sweep. Standard fittings for this use include waste T-fittings and 90° elbows. Y-fittings and 45° and 22° elbows can also be used for this purpose.

Vertical-to-horizontal drains: To change directions from the vertical to the horizontal, use fittings with a very pronounced, gradual sweep. Common fittings for this purpose include the combination Y-fitting with 45° elbow (often called a *combo*), and long sweep 90° elbow.

Horizontal offsets in drains: Y-fittings, 45° elbows, 22° elbows, and long sweep 90° elbows are used when changing directions in horizontal pipe runs. Whenever possible, horizontal drain pipes should use gradual, sweeping bends rather than sharp turns.

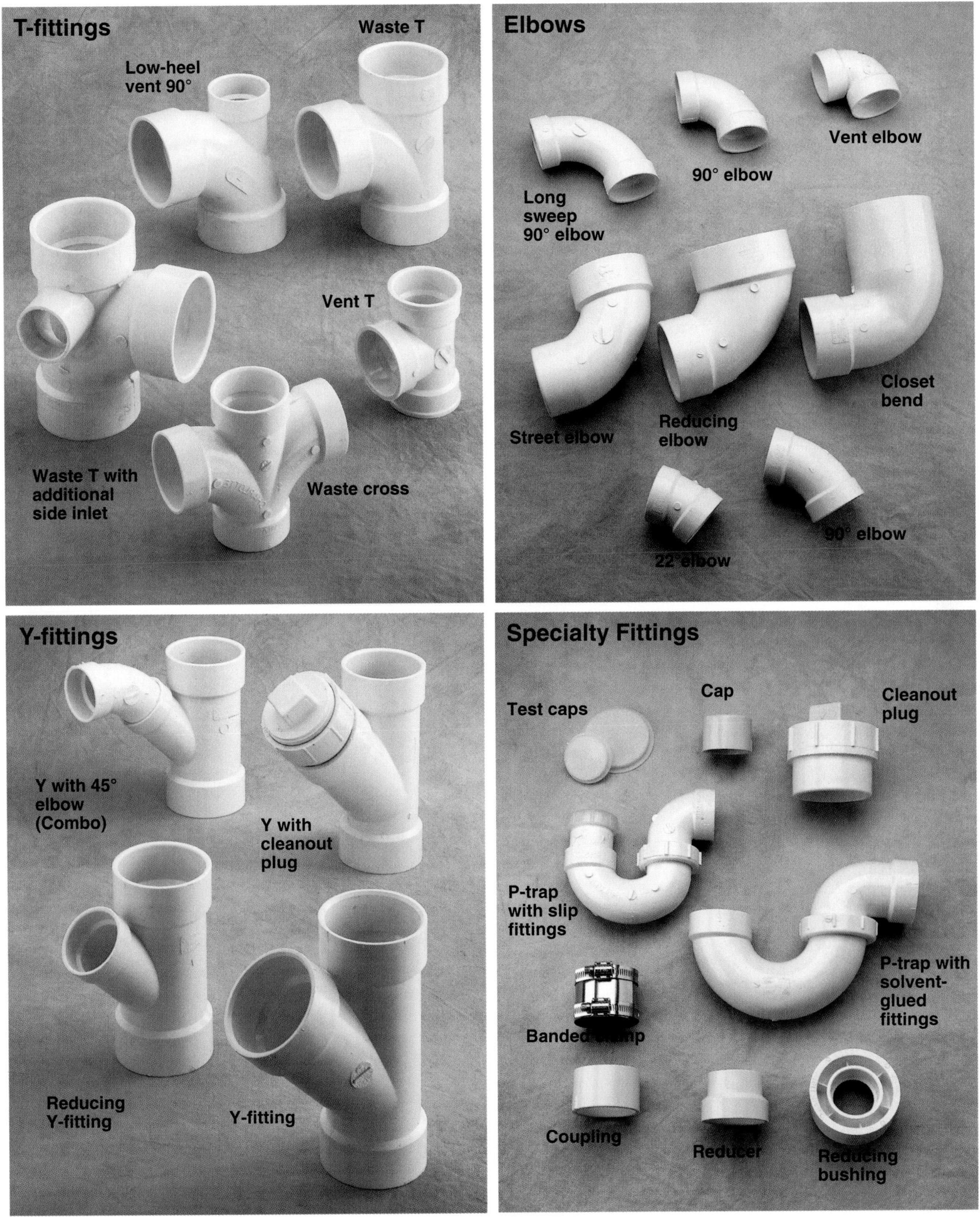

Fittings for DWV pipes are available in many configurations, with openings ranging from 1¼" to 4" in diameter. When planning your project, buy plentiful numbers of DWV and water supply fittings from a reputable retailer with a good return policy. It is much more efficient to return leftover materials after you complete your project than it is to interrupt your work each time you need to shop for a missing fitting.

Copper pipes and fittings are usually joined by soldering. First, the pipe and fitting are heated with a torch, then a length of metal soldering wire is touched to the joint, where it melts and is drawn up into the joint, sealing it.

Working with Copper Pipe

Copper pipes and fittings connected with soldered, or "sweated" joints are the professional standard for water supply pipes. Copper is more durable than CPVC plastic supply pipes, and is easier to fit and less prone to corrosion than threaded steel pipes. There are several grades of copper pipe sold, but for residential homes, ½"- and ¾"-diameter rigid type-L copper is the best choice.

Soldering copper pipe and fittings is a skill that requires some practice. Practice your soldering techniques on pieces of scrap pipe before you begin your project.

To ensure successful joints, it is essential that copper pipes and fittings be clean. Clean the ends of the pipes by sanding them with emery paper, and scour the insides of the fittings with a wire fitting brush before you solder. After the joint is completed, make sure not to disturb the pipes until the solder has lost its shiny color.

How to Cut & Join Copper Pipe

1 Measure and mark pipes to length, including the portion that will extend inside fittings. Place the tubing cutter over the pipe and tighten the handle until the cutting wheel is tight against the length mark. Rotate the cutter several times around the pipe, tightening the handle slightly with each turn, until the pipe is severed.

2 Use the reaming tool on the pipe cutter to remove the burr from the inside of the pipe. Rotate the reaming tool inside the pipe, using just enough pressure to remove the burr without bending the pipe.

3 Clean the inside of each fitting by scouring with a fitting brush. Apply a thin layer of soldering paste to the end of each pipe, using a flux brush, then join the pipes and fittings.

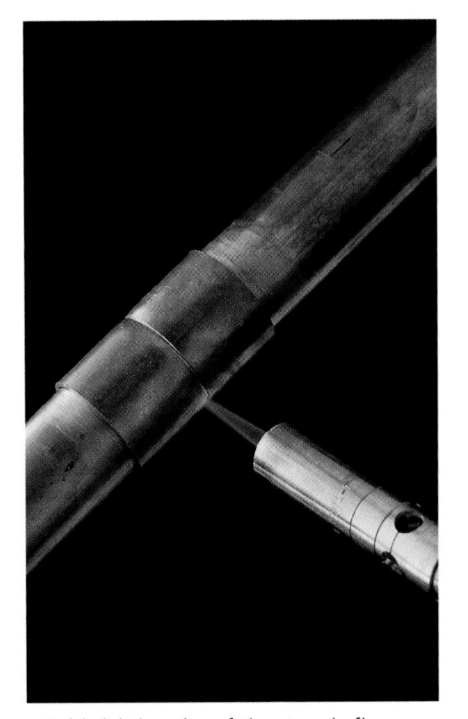

4 Hold the tip of the torch flame against the fitting until the soldering paste begins to sizzle, then move the flame around to the other side of the fitting to ensure even heat.

5 Touch the end of the solder to the pipe just below the fitting. If the solder melts, the pipe is sufficiently hot.

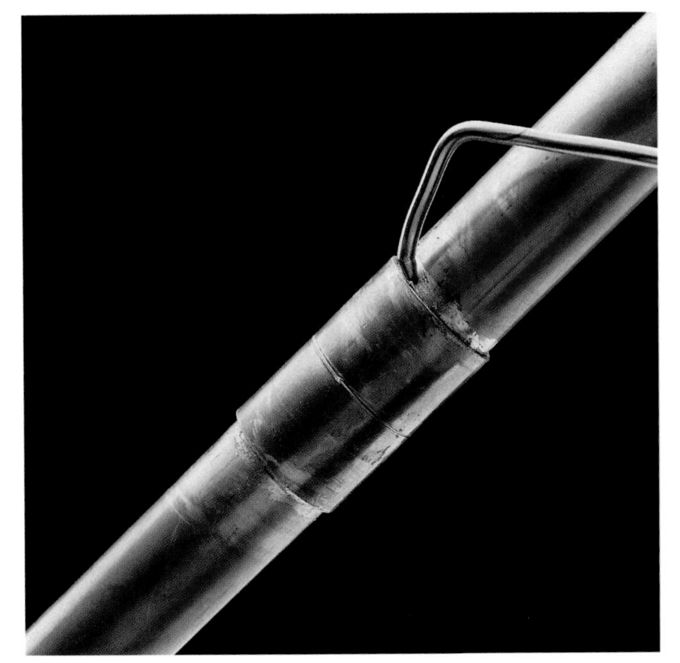

6 Quickly apply solder along both seams of the fitting, allowing the liquefied solder to be drawn into the fitting by capillary action. When the joint is filled, solder will begin to form droplets on the bottom of the joint. When correctly soldered, the joint will show a thin bead of silver-colored solder around each seam.

7 Allow the joint to cool until the solder has a frosty appearance rather than a shiny silver color, then wipe away any excess soldering paste with a wet cloth. Use caution: the pipes will be hot and may cause the damp cloth to steam.

Working with Rigid Plastic Pipe

Cut rigid plastic pipe to length, using a miter saw or handsaw designed for plastic. If you cut pipe with a handsaw, make sure the cuts are square. Use a utility knife to remove burrs from the cut edges of the pipe.

Rigid plastic pipe has largely replaced metal pipe in DWV installations. Most plumbers prefer PVC plastic over ABS, because it is sturdier and less flammable. CPVC is used for water supply pipe in some areas.

Rigid plastic pipes and fittings are joined with solvent glue. Make sure to use a glue made for the type of plastic you are installing. Some solvent glues, called "universal" solvents, can be used with all three types of rigid plastic pipe. Solvent-glued joints harden in about 30 seconds, so make sure to test-fit all joints before gluing them.

Materials for joining plastic pipes and fittings include solvent glue and primer with purple dye. Do not use clear primer, because the inspector will look for the purple stain as proof the pipes were primed. Liquid solvent glues and primers are toxic and flammable. Provide adequate ventilation when fitting plastics, and store the products always from any source of heat. Avoid getting primers and solvent-glue on your skin; many plumbers wear latex gloves to protect their skin.

How to Cut & Fit Rigid Plastic Pipe

1 Clean the surfaces to be joined by lightly scouring the ends of the pipe with emery paper, then wiping with a clean cloth. Cleaning the surfaces ensures that the primer will adequately soften the plastic before you apply solvent glue.

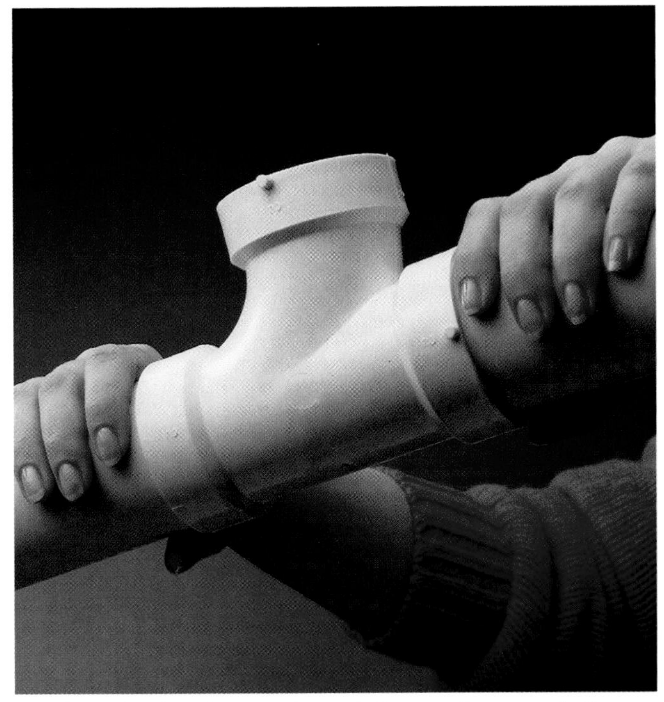

2 Cut plastic pipes to length (opposite page), then test-fit the pipes and fittings. Pipes should fit tightly against the bottom of the hubbed sockets on the fittings.

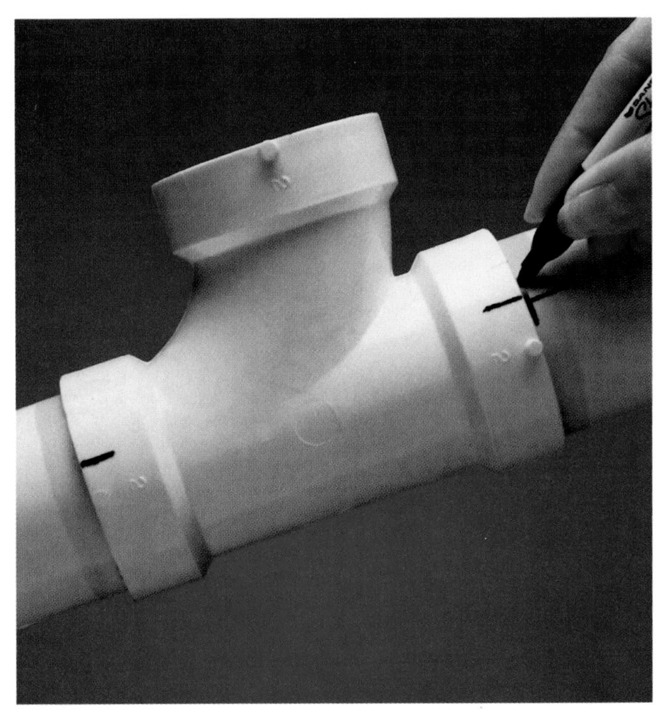

3 Make alignment marks across each joint with a felt-tipped pen, and mark the depth of the fitting sockets on the pipes. Take the pipes apart, and apply plastic primer to the surfaces being joined.

4 While the primer is still wet, apply a liberal layer of solvent glue to the end of the pipe and a thinner coat to the inside of the fitting socket. Work quickly, because the solvent glue will begin to dry within 30 seconds.

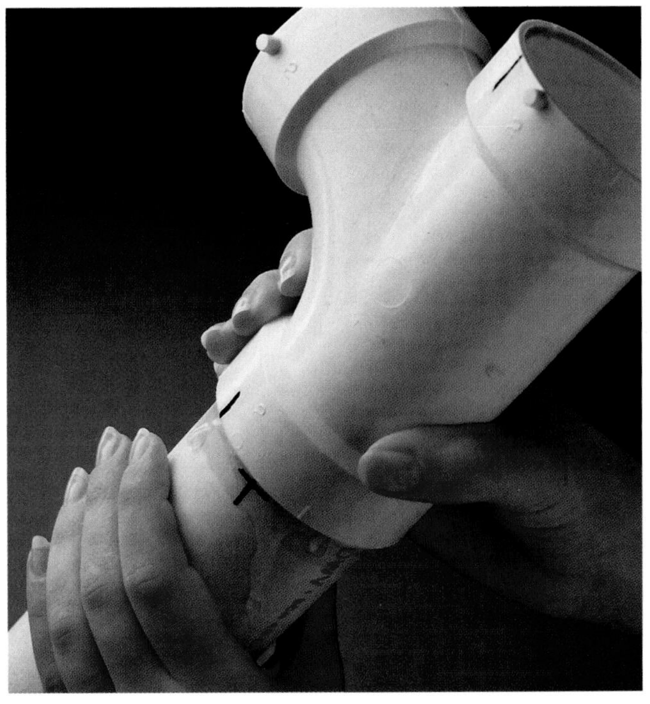

5 Quickly position pipe and fitting so that alignment marks are offset by about 2 inches. Force the pipe into the fitting until the end fits flush against the bottom of the socket, then twist the pipe into alignment. After completing the joint, wipe away the excess solvent with a rag.

Working with Flexible Plastic Pipe

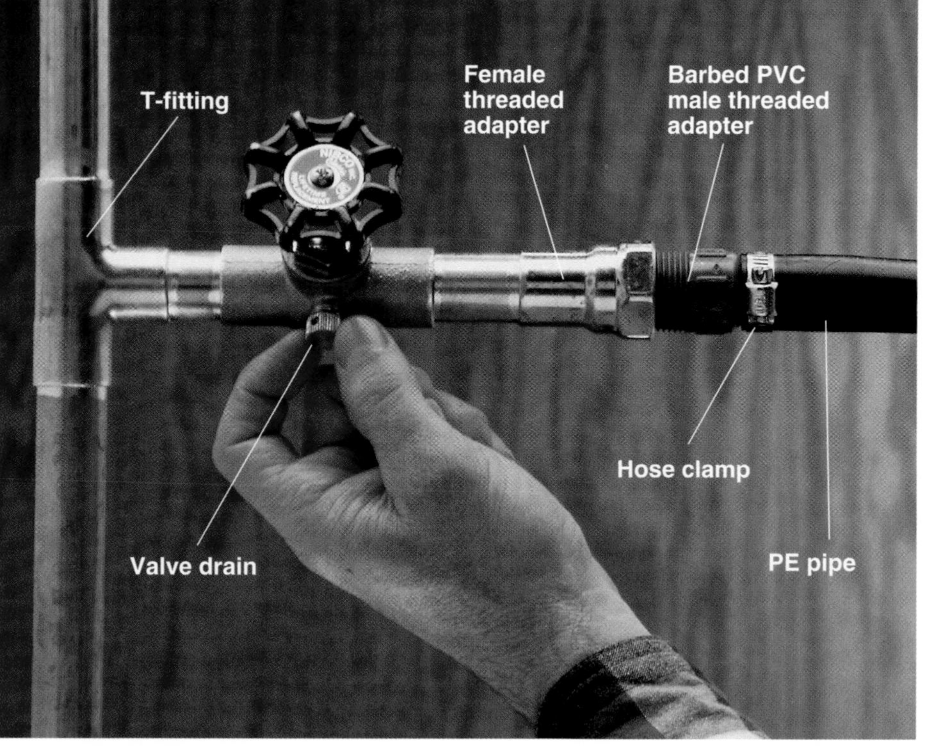

Connect PE pipe to an existing cold water supply pipe by splicing in a T-fitting to the copper pipe and attaching a drain-and-waste shutoff valve and a female threaded adapter. Screw a barbed PVC male threaded adapter into the copper fitting, then attach the PE pipe. The drain-and-waste valve allows you to blow the PE line free of water when winterizing the system.

Flexible PE (polyethylene) pipe is used for underground cold water lines. Very inexpensive, PE pipe is commonly used for automatic lawn sprinkler systems and for extending cold water supply to utility sinks in detached garages and sheds.

Unlike other plastics, PE is not solvent-glued, but is joined using "barbed" rigid PVC fittings and stainless steel hose clamps. In cold climates, outdoor plumbing lines should be shut off and drained for winter.

How to Cut & Join PE Pipe

1 Use a ratchet-style plastic pipe cutter or a miter saw to cut PE pipe to length. If you saw PE pipe, make sure to remove burrs from the cut ends, using a utility knife.

2 Use barbed PVC fittings to connect lengths of PE pipe. First, slide stainless steel hose clamps over the pipe, then force the ends of the pipe over the barbed portion of the fitting. Slide the clamps to the ends of the pipe, then tighten securely with a wrench or screwdriver. PVC fittings are available in many shapes; the fitting shown here has a threaded T that can be used to drain the pipes.

Working with Steel Pipe

Galvanized steel pipe is rarely used in new plumbing, but in the process of installing new water supply pipes you may need to remove steel pipe, or connect new copper pipes to existing steel pipes.

A water supply system made with steel pipe will have a large number of threaded fittings, and can be difficult if not impossible to disassemble. Most professional plumbers leave old water supply pipes in place, removing only the sections that interfere with the routing of new copper pipes.

Copper-to-steel transitions must be made with a dielectric union, a special fitting which keeps the dissimilar metals separate and prevents them from reacting with one another. Never thread copper and steel pipe directly together, because electrochemical reaction will cause the metals to corrode.

How to Remove Galvanized Pipe & Create a Transition

1 Cut through galvanized pipe with a reciprocating saw and metal-cutting blade, or with a hacksaw.

2 Hold the fitting with one pipe wrench, and use another wrench to remove the old pipe. The jaws of the wrenches should face opposite directions.

3 If you are attaching copper to the galvanized pipe, attach a dielectric union to create the transition. The union is soldered to the copper pipe and is threaded onto the galvanized steel.

Build a 2 x 4 support frame if a cast-iron waste-vent stack cannot be supported by joists—a situation that may occur in an attic. Blocking attached across the frame supports the riser clamps that grip the stack above and below the cutout section.

Working with Cast-iron Pipe

Cast-iron drain pipe is rarely installed these days, but if your home is more than 30 years old, there is a good chance you will encounter cast-iron pipe when renovating your plumbing system.

When installing new plastic DWV pipes, you may need to cut into a cast-iron waste-vent stack in order to connect the new drain and vent pipes. If the iron stack is in poor condition, you may need to replace it entirely with a new plastic waste-vent stack. A special fitting called a *banded coupling* is used to connect new plastic pipe to existing cast iron. Banded couplings come in several styles, so check with local Code to determine the correct fitting to use.

Cast iron is difficult to cut, and very heavy—an iron main waste-vent stack can weigh several hundred pounds. Always have a helper when working with cast-iron pipe, and make sure it is adequately supported before you cut into it.

Cast iron is best cut with a rented tool called a *snap cutter*. Snap cutter designs vary, so follow the rental dealer's instructions for using the tool.

How to Connect Plastic Pipe to Cast Iron

1 Use a piece of chalk to mark the cast iron for cutting. Make sure to remove enough pipe to accommodate the plastic fitting, pipes, and couplings needed to make the transition.

2 Support the lower section of the cast-iron pipe by installing a riser clamp flush against the bottom plate of the wall. Tighten the clamp firmly around the cast-iron waste-vent stack.

3 Support the upper section of the cast-iron pipe by installing a riser clamp 6" to 12" above the section of pipe to be removed. Attach wood blocks to the studs with 2½" wallboard screws, so that the riser clamp rests on the tops of the blocks, then tighten the clamp securely.

4 Wrap the chain of the snap cutter around the pipe, so that the cutting wheels are against the chalk outline. Tighten the chain and snap the pipe, following the tool manufacturer's directions.

5 Make a second cut at the other chalk line, then remove the cut section of pipe. Solvent-glue short lengths of plastic pipe to the sockets on the plastic fitting, so that the overall length of the new assembly is ½" shorter than cutout section.

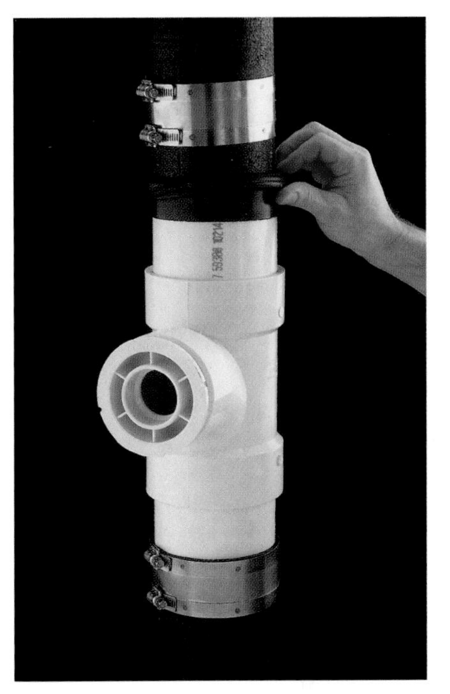

6 Slide a banded coupling onto each cut end of the cast-iron pipe, then roll back the rubber sleeves and insert plastic assembly. Roll the sleeves down onto the plastic pipes, then slide the metal bands in place and tighten them.

The plumbing inspector is the final authority when it comes to evaluating your work. By visually examining and testing your new plumbing, the inspector ensures that your work is safe and functional.

Understanding Plumbing Codes

The Plumbing Code is the set of regulations that building officials and inspectors use to evaluate your project plans and the quality of your work. Codes vary from region to region, but most are based on the National Uniform Plumbing Code, the authority we used in the development of this book.

Code books are available for reference at bookstores and government offices. However, they are highly technical, difficult-to-read manuals. More user-friendly for do-it-yourselfers are the variety of Code Handbooks available at bookstores and libraries. These handbooks are based on the National Uniform Plumbing Code, but are easier to read and include many helpful diagrams and photos.

Plumbing Code Handbooks sometimes discuss three different plumbing "zones" in an effort to accommodate variations in regulations from state to state. The states included in each zone are listed below.

Zone 1: Washington, Oregon, California, Nevada, Idaho, Montana, Wyoming, North Dakota, South Dakota, Minnesota, Iowa, Nebraska, Kansas, Utah, Arizona, Colorado, New Mexico, Indiana, parts of Texas.

Zone 2: Alabama, Arkansas, Louisiana, Tennessee, North Carolina, Mississippi, Georgia, Florida, South Carolina, parts of Texas, parts of Maryland, parts of Delaware, parts of Oklahoma, parts of West Virginia.

Zone 3: Virginia, Kentucky, Missouri, Illinois, Michigan, Ohio, Pennsylvania, New York, Connecticut, Massachusetts, Vermont, New Hampshire, Rhode Island, New Jersey, parts of Delaware, parts of West Virginia, parts of Maine, parts of Maryland, parts of Oklahoma.

Remember that your local Plumbing Code always supercedes the National Code. On some issues, the local Code may be less demanding than the National Code, but on other issues it may be more restrictive. Your local building inspector is a valuable source of information and may provide you with a convenient summary sheet of the regulations that apply to your project.

Getting a Permit

To ensure public safety, your community requires that you obtain a permit for most plumbing projects, including all the projects demonstrated in this book.

When you visit your city Building Inspection office to apply for a permit, the building official will want to review three drawings of your plumbing project: a site plan, a water supply diagram, and a drain-waste-vent diagram. These drawings are described on this page. If the official is satisfied that your project meets Code requirements, he will issue you a plumbing permit, which is your legal permission to begin work. The building official will also specify an inspection schedule for your project. As your project nears completion, you will be asked to arrange for an inspector to visit your home while the pipes are exposed and review the installation to ensure its safety.

Although do-it-yourselfers often complete complex plumbing projects without obtaining a permit or having the work inspected, we strongly urge you to comply with the legal requirements in your area. A flawed plumbing system can be dangerous, and it can potentially threaten the value of your home.

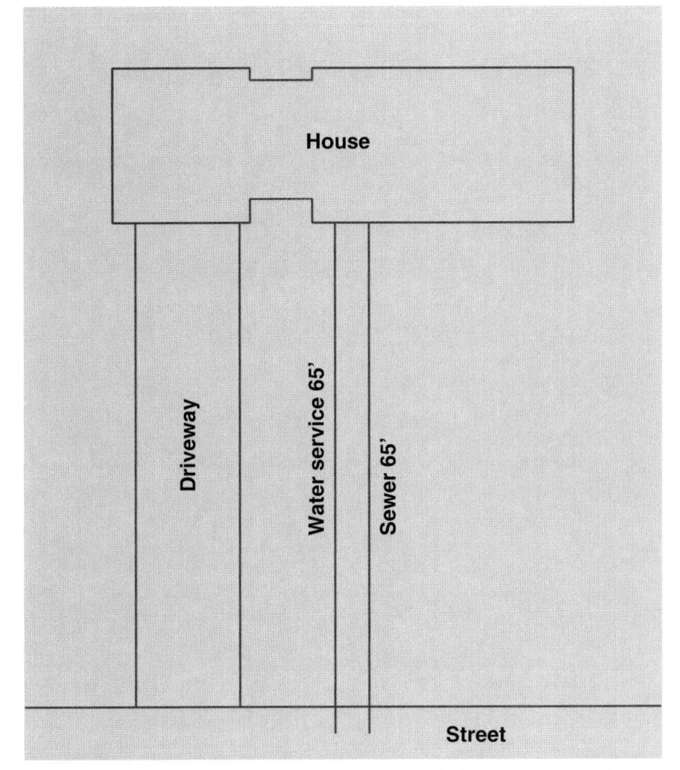

The site plan shows the location of the water main and sewer main with respect to your yard and home. The distances from your foundation to the water main and from the foundation to the main sewer should be indicated on the site plan.

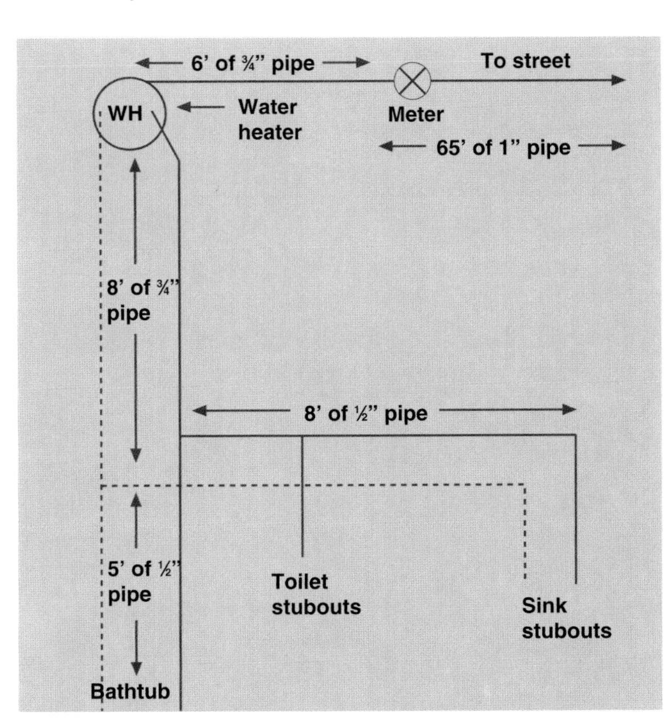

The supply riser diagram shows the length of the hot and cold water pipes and the relation of the fixtures to one another. The inspector will use this diagram to determine the proper size for the new water supply pipes in your system.

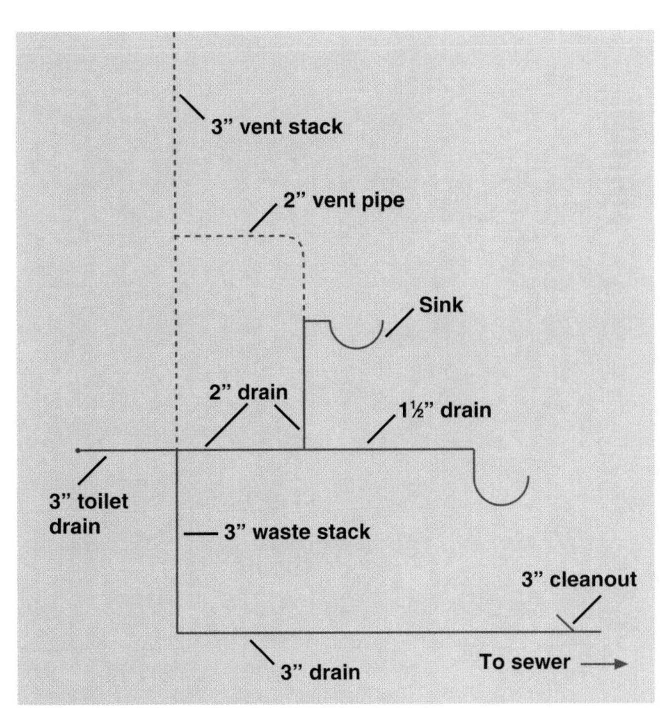

A DWV diagram shows the routing of drain and vent pipes in your system. Make sure to indicate the lengths of drain pipes and the distances between fixtures. The inspector will use this diagram to determine if you have properly sized the drain traps, drain pipes, and vent pipes in your project.

Sizing for Water Distribution Pipes

Fixture	Unit rating
Toilet	3
Vanity sink	1
Shower	2
Bathtub	2
Dishwasher	2
Kitchen sink	2
Clothes washer	2
Utility sink	2
Sillcock	3

Size of service pipe from street	Size of distribution pipe from water meter	Maximum length (ft.)—total fixture units					
		40	60	80	100	150	200
¾"	½"	9	8	7	6	5	4
¾"	¾"	27	23	19	17	14	11
¾"	1"	44	40	36	33	28	23
1"	1"	60	47	41	36	30	25
1"	1¼"	102	87	76	67	52	44

Water distribution pipes are the main pipes extending from the water meter throughout the house, supplying water to the branch pipes leading to individual fixtures. To determine the size of the distribution pipes, you must first calculate the total demand in "fixture units" (above, left) and the overall length of the water supply lines, from the street hookup through the water meter and to the most distant fixture in the house. Then, use the second table (above, right) to calculate the minimum size for the water distribution pipes. Note that the fixture unit capacity depends partly on the size of the street-side pipe that delivers water to your meter.

Sizes for Branch Pipes & Supply Tubes

Fixture	Min. branch pipe size	Min. supply tube size
Toilet	½"	⅜"
Vanity sink	½"	⅜"
Shower	½"	½"
Bathtub	½"	½"
Dishwasher	½"	½"
Kitchen sink	½"	½"
Clothes washer	½"	½"
Utility sink	½"	½"
Sillcock	¾"	N.A.
Water heater	¾"	N.A.

Branch pipes are the water supply lines that run from the distribution pipes toward the individual fixtures. **Supply tubes** are the vinyl, chromed copper, or mesh tubes that carry water from the branch pipes to the fixtures. Use the chart above as a guide when sizing branch pipes and supply tubes.

Valve Requirements

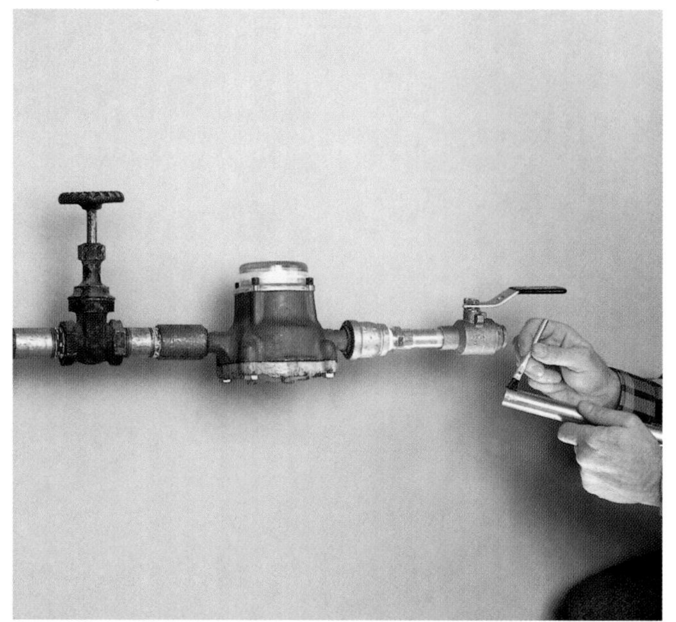

Full-bore gate valves or ball valves are required in the following locations: on both the street side and house side of the water meter; on the inlet pipes for water heaters and heating system boilers. Individual fixtures should have accessible shutoff valves, but these need not be full-bore valves. All sillcocks must have individual control valves located inside the house.

Modifying Water Pressure

Pressure-reducing valve (shown above) is required if the water pressure coming into your home is greater than 80 pounds per square inch (psi). The reducing valve should be installed near the point where the water service enters the building. A **booster pump** may be required if the water pressure in your home is below 40 psi.

Preventing Water Hammer

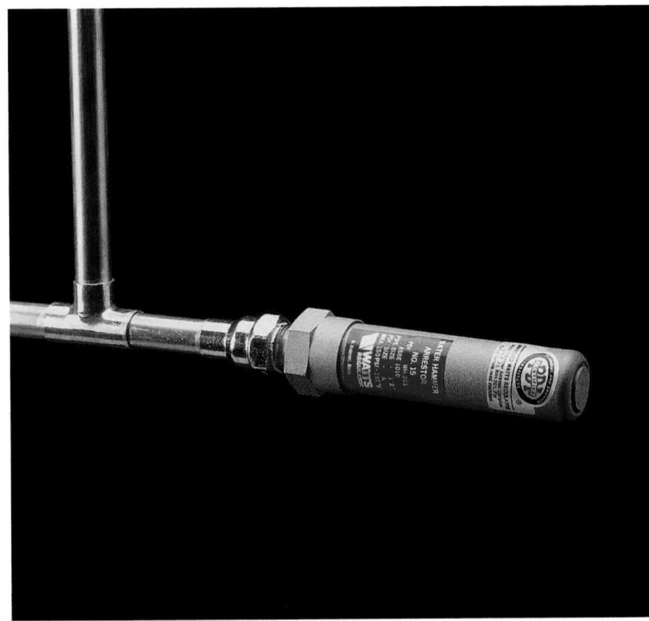

Water hammer arresters may be required by Code. Water hammer is a problem that may occur when the fast-acting valves on washing machines or other appliances cause pipes to vibrate against framing members. The arrester works as a shock absorber, with a watertight diaphragm inside. It is mounted to a T-fitting installed near the appliance.

Anti-syphon Devices

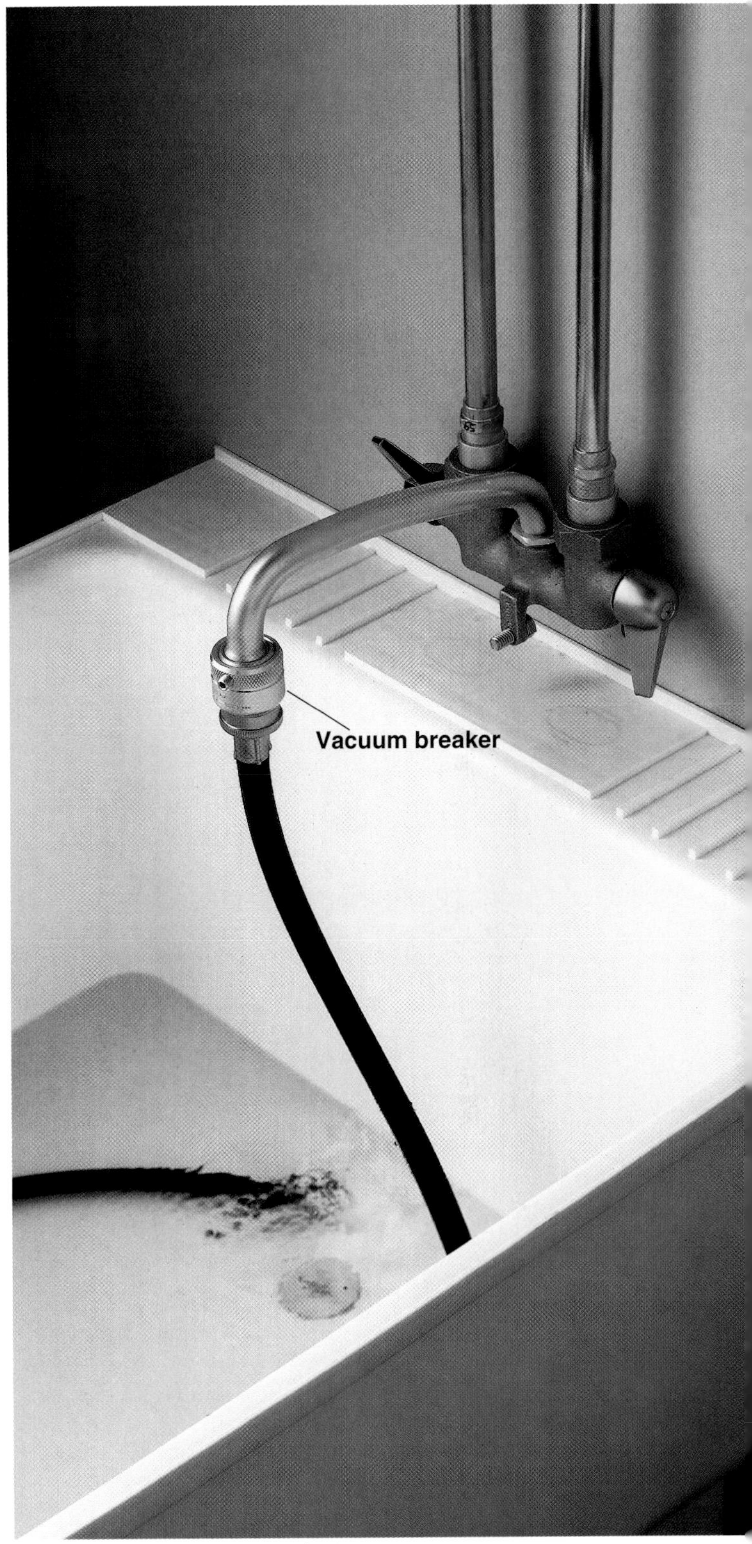

Vacuum breakers must be installed on all indoor and outdoor hose bibs and any outdoor branch pipes that run underground (page 81, step 7). Vacuum breakers prevent contaminated water from being drawn into the water supply pipes in the event of a sudden drop in water pressure in the water main. When a drop in pressure produces a partial vacuum, the breaker prevents siphoning by allowing air to enter the pipes.

Drain cleanouts make your DWV system easier to service. In most areas, the plumbing Code requires that you place cleanouts at the end of every horizontal drain run. Where horizontal runs are not accessible, removable drain traps will suffice as cleanouts.

Pipe Support Intervals

Type of pipe	Vertical support interval	Horizontal support interval
Copper	6 ft.	10 ft.
ABS	4 ft.	4 ft.
CPVC	3 ft.	3 ft.
PVC	4 ft.	4 ft.
Steel	12 ft.	15 ft.
Iron	5 ft.	15 ft.

Minimum intervals for supporting pipes are determined by the type of pipe and its orientation in the system. See page 23 for acceptable pipe support materials. Remember that the measurements shown above are minimum requirements; many plumbers install pipe supports at closer intervals.

Fixture Units & Minimum Trap Size

Fixture	Fixture units	Min. trap size
Shower	2	2"
Vanity sink	1	1¼"
Bathtub	2	1½"
Dishwasher	2	1½"
Kitchen sink	2	1½"
Kitchen sink*	3	1½"
Clothes washer	2	1½"
Utility sink	2	1½"
Floor drain	1	2"

*Kitchen sink with attached food disposer

Minimum trap size for fixtures is determined by the drain fixture unit rating, a unit of measure assigned by the Plumbing Code. NOTE: Kitchen sinks rate 3 units if they include an attached food disposer, 2 units otherwise.

Sizes for Horizontal & Vertical Drain Pipes

Pipe size	Maximum fixture units for horizontal branch drain	Maximum fixture units for vertical drain stacks
1¼"	1	2
1½"	3	4
2"	6	10
2½"	12	20
3"	20	30
4"	160	240

Drain pipe sizes are determined by the load on the pipes, as measured by the total fixture units. Horizontal drain pipes less than 3" in diameter should slope ¼" per foot toward the main drain. Pipes 3" or more in diameter should slope ⅛" per foot. NOTE: Horizontal or vertical drain pipes for a toilet must be 3" or larger.

Vent Pipe Sizes, Critical Distances

Size of fixture drain	Minimum vent pipe size	Maximum trap-to-vent distance
1¼"	1¼"	2½ ft.
1½"	1¼"	3½ ft.
2"	1½"	5 ft.
3"	2"	6 ft.
4"	3"	10 ft.

Vent pipes are usually one pipe size smaller than the drain pipes they serve. Code requires that the distance between the drain trap and the vent pipe fall within a maximum "critical distance," a measurement that is determined by the size of the fixture drain. Use this chart to determine both the minimum size for the vent pipe and the maximum critical distance.

Vent Pipe Orientation to Drain Pipe

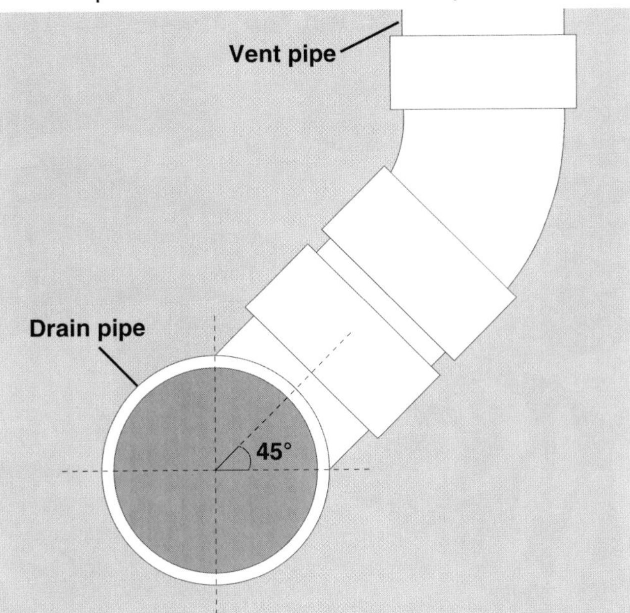

Vent pipes must extend in an upward direction from drains, no less than 45° from horizontal, This ensures that waste water cannot flow into the vent pipe and block it. At the opposite end, a new vent pipe should connect to an existing vent pipe or main waste-vent stack at a point at least 6" above the highest fixture draining into the system.

Wet Venting

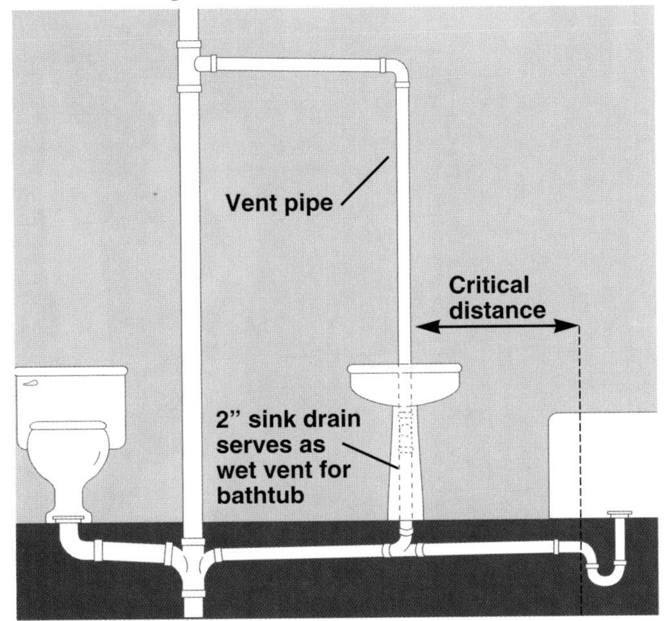

Wet vents are pipes that serve as a vent for one fixture and a drain for another. The sizing of a wet vent is based on the total fixture units it supports (opposite page): a 3" wet vent can serve up to 12 fixture units; a 2" wet vent is rated for 4 fixture units; a 1½" wet vent, for only 1 fixture unit. NOTE: The distance between the wet-vented fixture and the wet vent itself must be no more than the maximum critical distance (above, left).

Auxiliary Venting

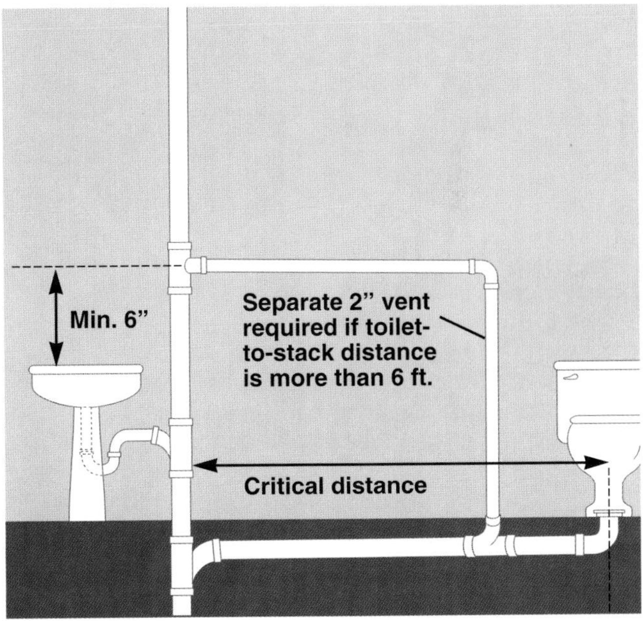

Fixtures must have auxiliary vents if the distance to the main waste-vent stack exceeds the critical distance (above, left). A toilet, for example, should have a separate vent pipe if it is located more than 6 ft. from the main waste-vent stack. This secondary vent pipe should connect to the stack or an existing vent pipe at a point at least 6" above the highest fixture on the system.

A pressure gauge and air pump are used to test DWV lines. The system is first blocked off at each fixture and at points near where the new drain and vent pipes connect to the main stack. Air is then pumped into the system to a pressure of 5 pounds per square inch (PSI). To pass inspection, the system must hold this pressure for 15 minutes.

Testing New Plumbing Pipes

When the building inspector comes to review your new plumbing, he may require that you perform a pressure test on the DWV and water supply lines as he watches. The inspection and test should be performed after the system is completed, but before the new pipes are covered with wallboard. To ensure that the inspection goes smoothly, it is a good idea to perform your own pretest, so you can locate and repair any problems before the inspector makes his visit.

The DWV system is tested by blocking off the new drain and vent pipes, then pressuring the system with air to see if it leaks. At the fixture stub-outs, the DWV pipes can be capped off or plugged with test balloons designed for this purpose. The air pump, pressure gauge, and test balloons required to test the DWV system can be obtained at tool rental centers.

Testing the water supply lines is a simple matter of turning on the water and examining the joints for leaks. If you find a leak, you will need to drain the pipes, then resolder the faulty joints.

How to Test New DWV Pipes

1 Insert a test balloon into the test T-fittings at the top and bottom of the new DWV line, blocking the pipes entirely. NOTE: Ordinary T-fittings installed near the bottom of the drain line and near the top of the vent line are generally used for test fittings.

2 Block toilet drains with a test balloon designed for a toilet bend. Large test balloons may need to be inflated with an air pump.

3 Cap off the remaining fixture drains by solvent-gluing test caps onto the stub-outs. After the system is tested, these caps are simply knocked loose with a hammer.

4 At a cleanout fitting, insert a *weenie*—a special test balloon with an air gauge and inflation valve. Attach an air pump to the valve on the weenie, and pressurize the pipes to 5 psi. Watch the pressure gauge for 15 minutes to ensure that the system does not lose pressure.

5 If the DWV system loses air when pressurized, check each joint for leaks by rubbing soapy water over the fittings and looking for active bubbles. When you identify a problem joint, cut away the existing fitting and solvent-glue a new fitting in place, using couplings and short lengths of pipe.

6 After the DWV system has been inspected and approved by a building official, remove the test balloons and close the test T-fittings by solvent-gluing caps onto the open inlets.

New Installation

BLACK & DECKER
INDUSTRY & CONSTRUCTION

Use 2 x 6 studs to frame "wet walls" when constructing a new bathroom or kitchen. Thicker walls provide more room to run drain pipes and main waste-vent stacks, making installation much easier.

Installing New Plumbing

A major plumbing project is a complicated affair that often requires demolition and carpentry skills. Bathroom or kitchen plumbing may be unusable for several days while completing the work, so make sure you have a backup bathroom or kitchen space to use during this time.

To ensure that your project goes quickly, always buy plenty of pipe and fittings—at least 25% more than you think you need. Making several extra trips to the building center for last-minute fittings is a nuisance, and it can add many hours of time to your project. Always purchase from a reputable retailer that will allow you to return leftover fittings for credit.

The how-to projects on the following pages demonstrate standard plumbing techniques, but should not be used as a literal blueprint for your own work. Pipe and fitting sizes, fixture layout, and pipe routing will always vary according to individual circumstances. When planning your project, carefully read all the information in the Planning section, especially the material on Understanding Plumbing Codes (pages 34 to 39). Before you begin work, create a detailed plumbing plan to guide your work and help you obtain the required permits. This section includes information on:

- Plumbing Bathrooms (pages 48 to 65)
- Plumbing a Kitchen (pages 66 to 77)
- Installing Outdoor Plumbing (pages 78 to 83)

Tips for Installing New Plumbing

Use masking tape to mark the locations of fixtures and pipes on the walls and floors. Read the layout specifications that come with each sink, tub, or toilet, then mark the drain and supply lines accordingly. Position the fixtures on the floor, and outline them with tape. Measure and adjust until the arrangement is comfortable to you and meets minimum clearance specifications. If you are working in a finished room, prevent damage to wallpaper or paint by using self-adhesive notes to mark the walls.

Consider the location of cabinets when roughing in the water supply and drain stub-outs. You may want to temporarily position the cabinets in their final locations before completing the drain and water supply runs.

Install control valves at the points where the new branch supply lines meet the main distribution pipes. By installing valves, you can continue to supply the rest of the house with water while you are working on the new branches.

(continued next page)

Tips for Installing New Plumbing (continued)

Framing Member	Maximum Hole Size	Maximum Notch Size
2 x 4 loadbearing stud	1 7/16" diameter	7/8" deep
2 x 4 non-loadbearing stud	2 1/2" diameter	1 7/16" deep
2 x 6 loadbearing stud	2 1/4" diameter	1 3/8" deep
2 x 6 non-loadbearing stud	3 5/16" diameter	2 3/16" deep
2 x 6 joists	1 1/2" diameter	7/8" deep
2 x 8 joists	2 3/8" diameter	1 1/4" deep
2 x 10 joists	3 1/16" diameter	1 1/2" deep
2 x 12 joists	3 3/4" diameter	1 7/8" deep

Framing member chart shows the maximum sizes for holes and notches that can be cut into studs and joists when running pipes. Where possible, use notches rather than bored holes, because pipe installation is usually easer. When boring holes, there must be at least 5/8" of wood between the edge of a stud and the hole, and at least 2" between the edge of a joist and the hole. Joists can be notched only in the end one-third of the overall span; never in the middle one-third of the joist. When two pipes are run through a stud, the pipes should be stacked one over the other, never side by side.

Create access panels so that in the future you will be able to service fixture fittings and shutoff valves located inside the walls. Frame an opening between studs, then trim the opening with wood moldings. Cover the opening with a removable plywood panel the same thickness as the wall surface, then finish it to match the surrounding walls.

Protect pipes from punctures, if they are less than 1 1/4" from the front face of wall studs or joists, by attaching metal protector plates to the framing members.

Test-fit materials before solvent-gluing or soldering joints. Test-fitting ensures that you have the correct fittings and enough pipe to do the job, and can help you avoid lengthy delays during installation.

Support pipes adequately. Horizontal and vertical runs of DWV and water supply pipe must be supported at minimum intervals, which are specified by your local Plumbing Code (page 38). A variety of metal and plastic materials are available for supporting plumbing pipes (page 23).

Use plastic bushings to help hold plumbing pipes securely in holes bored through wall plates, studs, and joists. Bushings can help to cushion the pipes, preventing wear and reducing rattling.

Install extra T-fittings on new drain and vent lines so that you can pressure-test the system when the building inspector reviews your installation (pages 40 to 41). A new DWV line should have these extra T-fittings near the points where the new branch drains and vent pipes reach the main waste-vent stack.

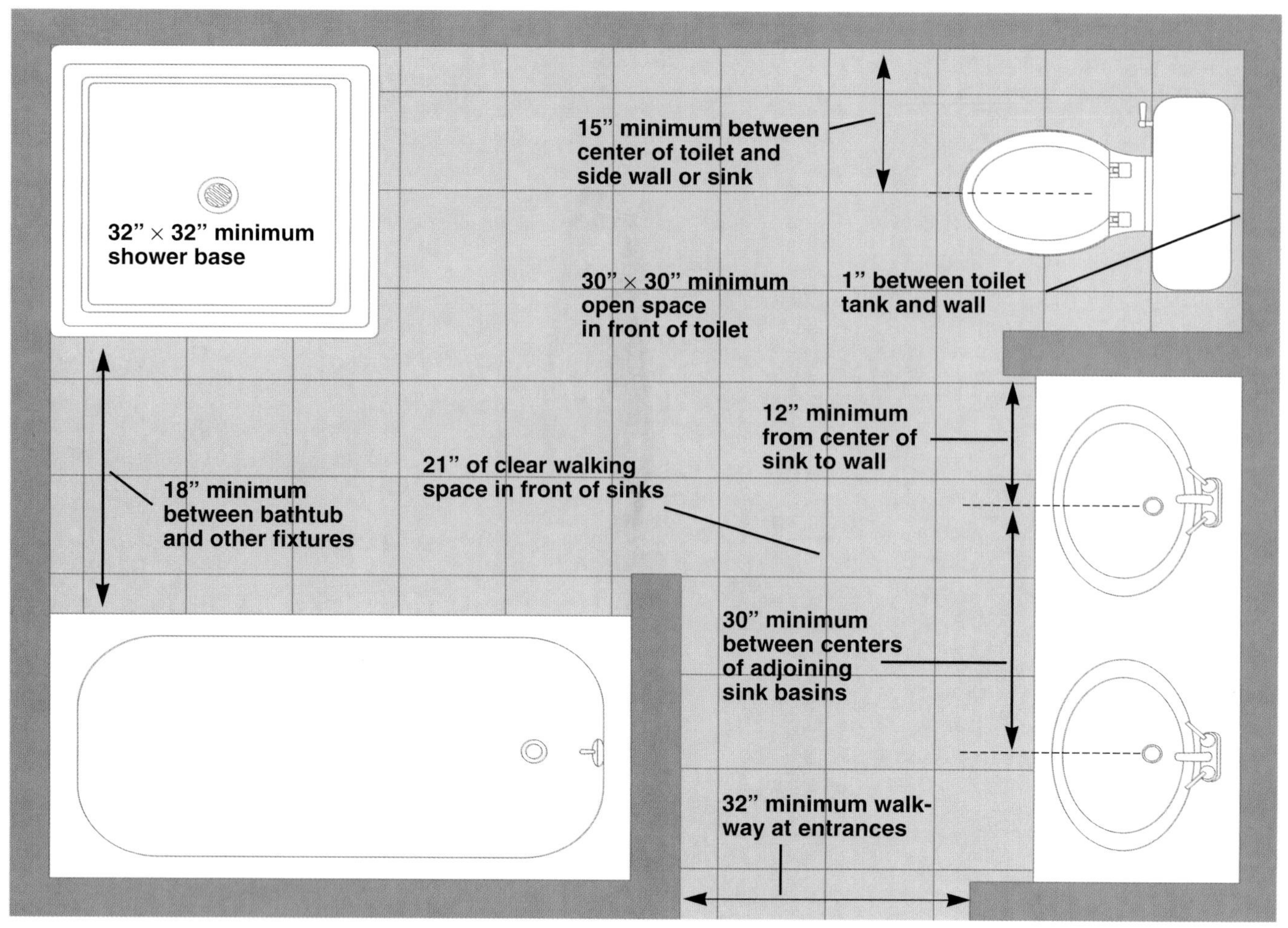

Follow minimum clearance guidelines when planning the locations of bathroom fixtures. Unobstructed access to the fixtures is fundamental to creating a comfortable, safe, and efficient bathroom.

Plumbing Bathrooms

Adding a new bathroom or updating an old one is a sure way to add value to your home. The personal comfort gained from a custom master bathroom can add a new dimension to your relaxation time. And adding a full bath in the basement or a half bath next to the kitchen offers convenience for both family members and guests. When planned and built correctly, a new or remodeled bathroom also improves the resale value of your home.

The first step when planning a bathroom is determining the type of bathroom you want. Do you have your heart set on expanding into a spare bedroom to create the ultimate master bathroom, or do you simply need a functional half bath for convenience? In this section, you'll see three demonstration projects that represent the full range of bathrooms, from a spacious master bathroom to a simple half bath (opposite page).

Next, you'll need to decide on the type of fixtures you need. Visit your local building centers to determine the range of fixtures available and their prices, and visit model homes and remodeling exhibitions to see how professionals arrange and install the fixtures you plan to use.

Once you have determined the scope of your project and settled on a budget, you can develop working plans for your bathroom. When creating a floor plan, always follow minimum clearance guidelines (above), and think about where the drain and water supply pipes will run. You can save yourself many hours of work by positioning fixtures so the pipes can be routed with simple, straight runs rather than with many complicated bends. Make sure that your project complies with local Plumbing Code regulations for bathroom plumbing.

Demonstration Bathroom Projects

Master bathroom can include luxury features, such as a large whirlpool tub or multi-jet shower. Our project contains both these features, as well as a pedestal vanity sink and toilet. A spacious bathroom may require substantial construction work if you intend to expand into an adjoining room. See pages 50 to 57.

Basement bathroom is ideal if you have bedrooms or finished recreation areas in your basement. Our project includes a shower, toilet, and vanity sink. Plumbing a basement bathroom may require that you break into the concrete floor to connect drain pipes. See pages 58 to 63.

Half bath can be easily added to a room that shares a "wet wall" with a kitchen or other bathroom. Our project includes a toilet and vanity sink. See pages 64 to 65.

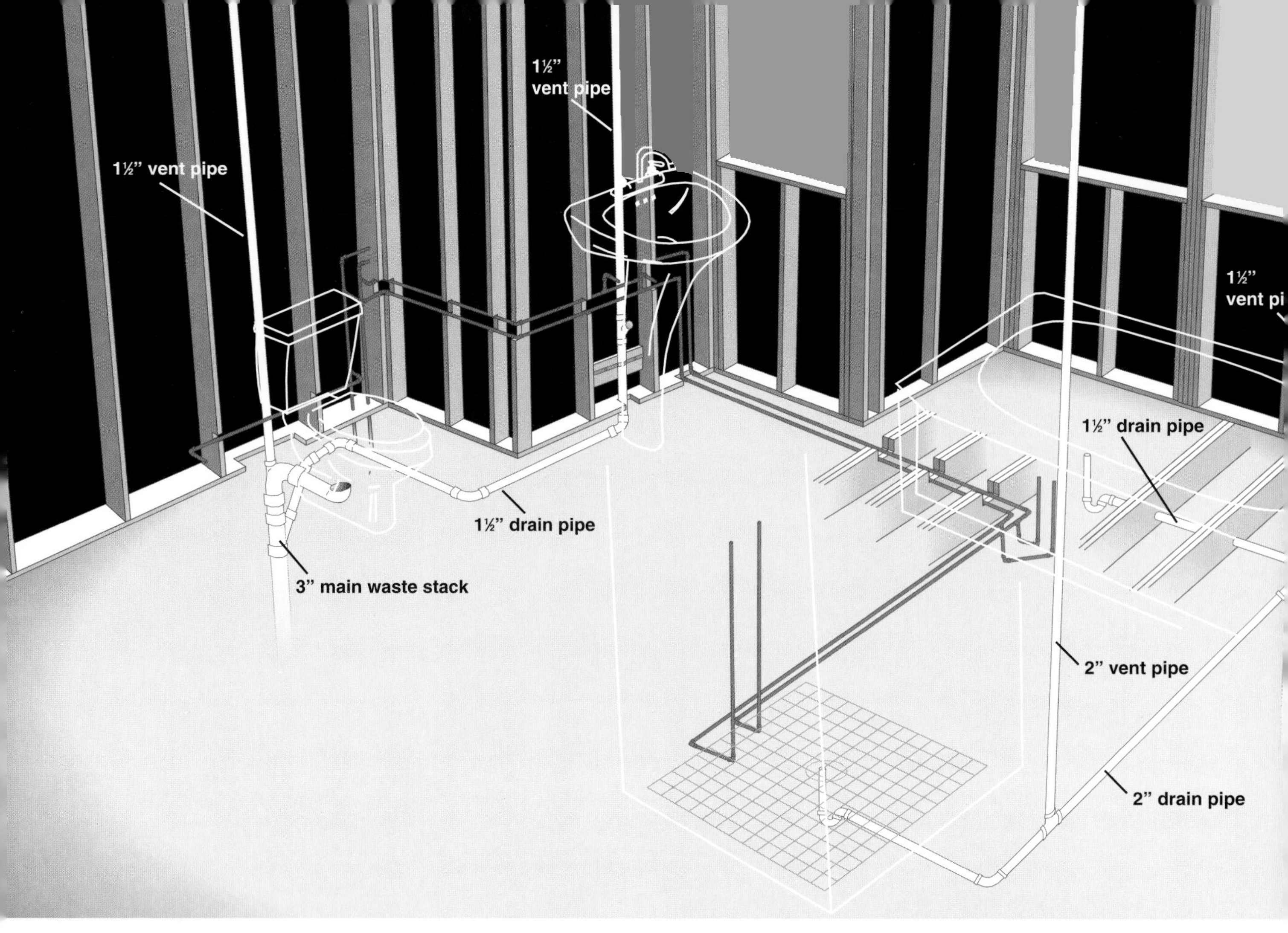

Plumbing a Master Bathroom

A large bathroom has more plumbing fixtures and consumes more water than any other room in your house. For this reason, a master bath has special plumbing needs.

Frame bathroom "wet walls" with 2 × 6 studs, to provide plenty of room for running 3" pipes and fittings. If your bathroom includes a heavy whirlpool tub, you will need to strengthen the floor by installing "sister" joists alongside the existing floor joists underneath the tub.

For convenience, our project is divided into the following sequences:

- How to Install DWV Pipes for the Toilet & Sink (pages 51 to 53)
- How to Install DWV Pipes for the Tub & Shower (pages 54 to 55)
- How to Connect the Drain Pipes & Vent Pipes to the Main Waste-Vent Stack (page 56)
- How to Install the Water Supply Pipes (page 57)

Our demonstration bathroom is a second-story master bath. We are installing a 3" vertical drain pipe to service the toilet and the vanity sink, and a 2" vertical pipe to handle the tub and shower drains. The branch drains for the sink and bathtub are 1½" pipes; for the shower, 2" pipe. Each fixture has its own vent pipe extending up into the attic, where they are joined together and connected to the main stack.

How to Install DWV Pipes for the Toilet & Sink

1 Use masking tape to outline the locations of the fixtures and pipe runs on the subfloor and walls. Mark the location for a 3" vertical drain pipe on the sole plate in the wall behind the toilet. Mark a 4½"-diameter circle for the toilet drain on the subfloor.

2 Cut out the drain opening for the toilet, using a jig saw. Mark and remove a section of flooring around the toilet area, large enough to provide access for installing the toilet drain and for running drain pipe from the sink. Use a circular saw with blade set to the thickness of the flooring to cut through the subfloor.

3 If a floor joist interferes with the toilet drain, cut away a short section of the joist and box-frame the area with double headers.The framed opening should be just large enough to install the toilet and sink drains.

4 To create a path for the vertical 3" drain pipe, cut a 4½" × 12" notch in the sole plate of the wall behind the toilet. Make a similar cutout in the double wall plate at the bottom of the joist cavity. From the basement, locate the point directly below the cutout by measuring from a reference point, such as the main waste-vent stack.

5 Mark the location for the 3" drain pipe on the basement ceiling, then drill a 1"-diameter hole up through the center of the marked area. Direct the beam of a bright flashlight up into the hole, then return to the bathroom and look down into the wall cavity. If you can see light, return to the basement and cut a 4½"-diameter hole centered over the test hole.

(continued next page)

6 Measure and cut a length of 3" drain pipe to reach from the bathroom floor cavity to a point flush with the bottom of the ceiling joists in the basement. Solvent-glue a 3" × 3" × 1½" Y-fitting to the top of the pipe, and a low-heel vent 90° fitting above the Y. The branch inlet on the Y should face toward the sink location; the front inlet on the low-heel should face forward. Carefully lower the pipe into the wall cavity.

7 Lower the pipe so the bottom end slides through the opening in the basement ceiling. Support the pipe with vinyl pipe strap wrapped around the low-heel vent 90° fitting and screwed to framing members.

8 Use a length of 3" pipe and a 4" × 3" reducing elbow to extend the drain out to the toilet location. Make sure the drain slopes at least ⅛" per foot toward the wall, then support it with pipe strap attached to the joists. Insert a short length of pipe into the elbow, so it extends at least 2" above the subfloor. After the new drains are pressure tested, this stub-out will be cut flush with the subfloor and fitted with a toilet flange.

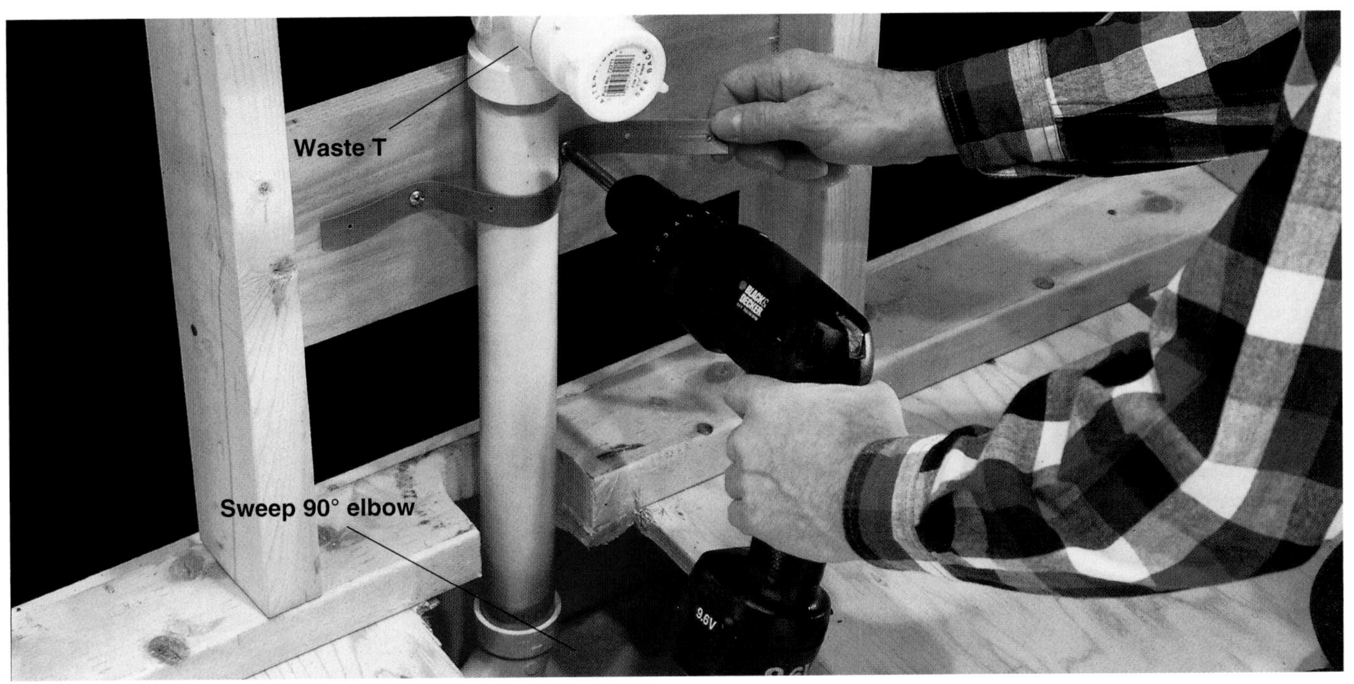

9 Notch out the sole plate and subfloor below the sink location. Cut a length of 1½" plastic drain pipe, then solvent-glue a waste T to the top of the pipe and a sweep 90° elbow to the bottom. The branch of the T should face out, and the discharge on the elbow should face toward the toilet location. Adjust the pipe so the top edge of the elbow nearly touches the bottom of the sole plate. Anchor it with a ¾"-thick backing board nailed between the studs.

10 Dry-fit lengths of 1½" drain pipe and elbows to extend the sink drain to the 3" drain pipe behind the toilet. Use a right-angle drill to bore holes in joists, if needed. Make sure the horizontal drain pipe slopes at least ¼" per foot toward the vertical drain. When satisfied with the layout, solvent-glue the pieces together and support the drain pipe with vinyl pipe straps attached to the joists.

11 In the top plates of the walls behind the sink and toilet, bore ½"-diameter holes up into the attic. Insert pencils or dowels into the holes, and tape them in place. Enter the attic and locate the pencils, then clear away insulation and cut 2"-diameter holes for the vertical vent pipes. Cut and install 1½" vent pipes running from the toilet and sink drain at least 1 ft. up into the attic.

How to Install DWV Pipes for the Tub & Shower

1 On the subfloor, use masking tape to mark the locations of the tub and shower, the water supply pipes, and the tub and shower drains, according to your plumbing plan. Use a jig saw to cut out a 12"-square opening for each drain, and drill 1"-diameter holes in the subfloor for each water supply riser.

2 When installing a large whirlpool tub, cut away the subfloor to expose the full length of the joists under the tub, then screw or bolt a second joist, called a *sister,* against each existing joist. Make sure both ends of each joist are supported by loadbearing walls.

3 In a wall adjacent to the tub, establish a route for a 2" vertical waste-vent pipe running from basement to attic. This pipe should be no more than 3½ ft. from the bathtub trap. Then, mark a route for the horizontal drain pipe running from the bathtub drain to the waste-vent pipe location. Cut 3"-diameter holes through the centers of the joists for the bathtub drain.

4 Cut and install a vertical 2" drain pipe running from basement to the joist cavity adjoining the tub location, using the same technique as for the toilet drain (steps 4 to 6, pages 51 to 52). At the top of the drain pipe, use assorted fittings to create three inlets: branch inlets for the bathtub and shower drains, and a 1½" top inlet for a vent pipe running to the attic.

5 Dry-fit a 1½" drain pipe running from the bathtub drain location to the vertical waste-vent pipe in the wall. Make sure the pipe slopes ¼" per foot toward the wall. When satisfied with the layout, solvent-glue the pieces together and support the pipe with vinyl pipe straps attached to the joists.

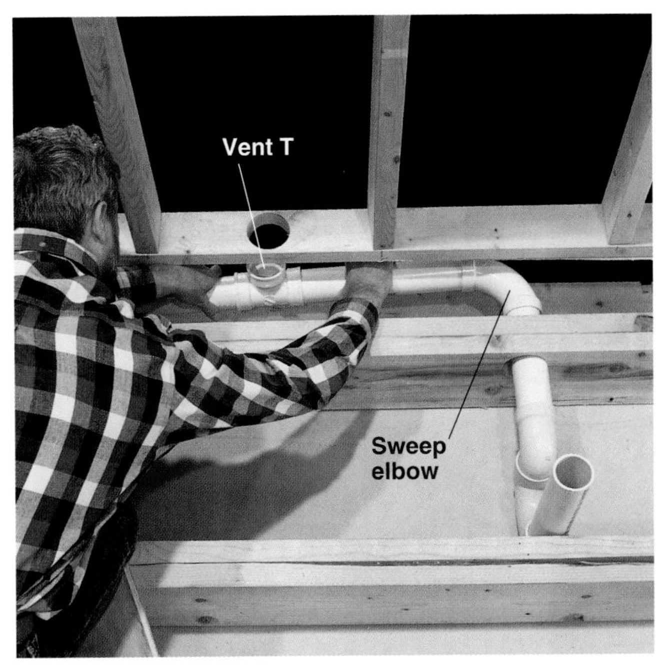

6 Dry-fit a 2" drain pipe from the shower drain to the vertical waste-vent pipe near the tub. Install a solvent-glued trap at the drain location, and cut a hole in the sole plate and insert a 2" × 2" × 1½" vent T within 5 ft. of the trap. Make sure the drain is sloped ¼" per foot downward away from the shower drain. When satisfied with the layout, solvent-glue the pipes together.

7 Cut and install vertical vent pipes for the bathtub and shower, extending up through the wall plates and at least 1 ft. into the attic. These vent pipes will be connected in the attic to the main waste-vent stack. In our project, the shower vent is a 2" pipe, while the bathtub vent is a 1½" pipe.

How to Connect the Drain Pipes to the Main Waste-Vent Stack

1 In the basement, cut into the main waste-vent stack and install the fittings necessary to connect the 3" toilet-sink drain and the 2" bathtub-shower drain. In our project, we created an assembly made of a waste T-fitting with an extra side inlet and two short lengths of pipe, then inserted it into the existing waste-vent stack using banded couplings (pages 32 to 33). Make sure the T-fittings are positioned so the drain pipes will have the proper downward slope toward the stack.

2 Dry-fit Y-fittings with 45° elbows onto the vertical 3" and 2" drain pipes. Position the horizontal drain pipes against the fittings, and mark them for cutting. When satisfied with the layout, solvent-glue the pipes together, then support the pipes every 4 ft. with vinyl pipe straps. Solvent-glue cleanout plugs on the open inlets on the Y-fittings.

How to Connect the Vent Pipes to the Main Waste-Vent Stack

1 In the attic, cut into the main waste-vent stack and install a vent T-fitting, using banded couplings. The side outlet on the vent T should face the new 2" vent pipe running down to the bathroom. Attach a test T-fitting to the vent T. NOTE: If your stack is cast iron, make sure to adequately support it before cutting into it (page 32).

2 Use elbows, vent T-fittings, reducers, and lengths of pipe as needed to link the new vent pipes to the test T-fitting on the main waste-vent stack. Vent pipes can be routed in many ways, but you should make sure the pipes have a slight downward angle to prevent moisture from collecting in the pipes. Support the pipes every 4 ft.

How to Install the Water Supply Pipes

1 After shutting off the water, cut into existing supply pipes and install T-fittings for new branch lines. Notch out studs and run copper pipes to the toilet and sink locations. Use an elbow and threaded female fitting to form the toilet stub-out. Once satisfied with the layout, solder the pipes in place.

2 Cut 1" × 4"-high notches around the wall, and extend the supply pipes to the sink location. Install reducing T-fittings and female threaded fittings for the sink faucet stub-outs. The stub-outs should be positioned about 18" above the floor, spaced 8" apart. Once satisfied with the layout, solder the joints, then insert ¾" blocking behind the stub-outs and strap them in place.

3 Extend the water supply pipes to the bathtub and shower. In our project, we removed the subfloor and notched the joists to run ¾" supply pipes from the sink to a whirlpool bathtub, then to the shower. At the bathtub, we used reducing T-fittings and elbows to create ½" risers for the tub faucet. Solder caps onto the risers; after the subfloor is replaced, the caps will be removed and replaced with shutoff valves.

4 At the shower location, use elbows to create vertical risers where the shower wet wall will be constructed. The risers should extend at least 6" above floor level. Support the risers with a ¾" backer board attached between joists. Solder caps onto the risers. After the shower stall is constructed, the caps will be removed and replaced with shutoff valves.

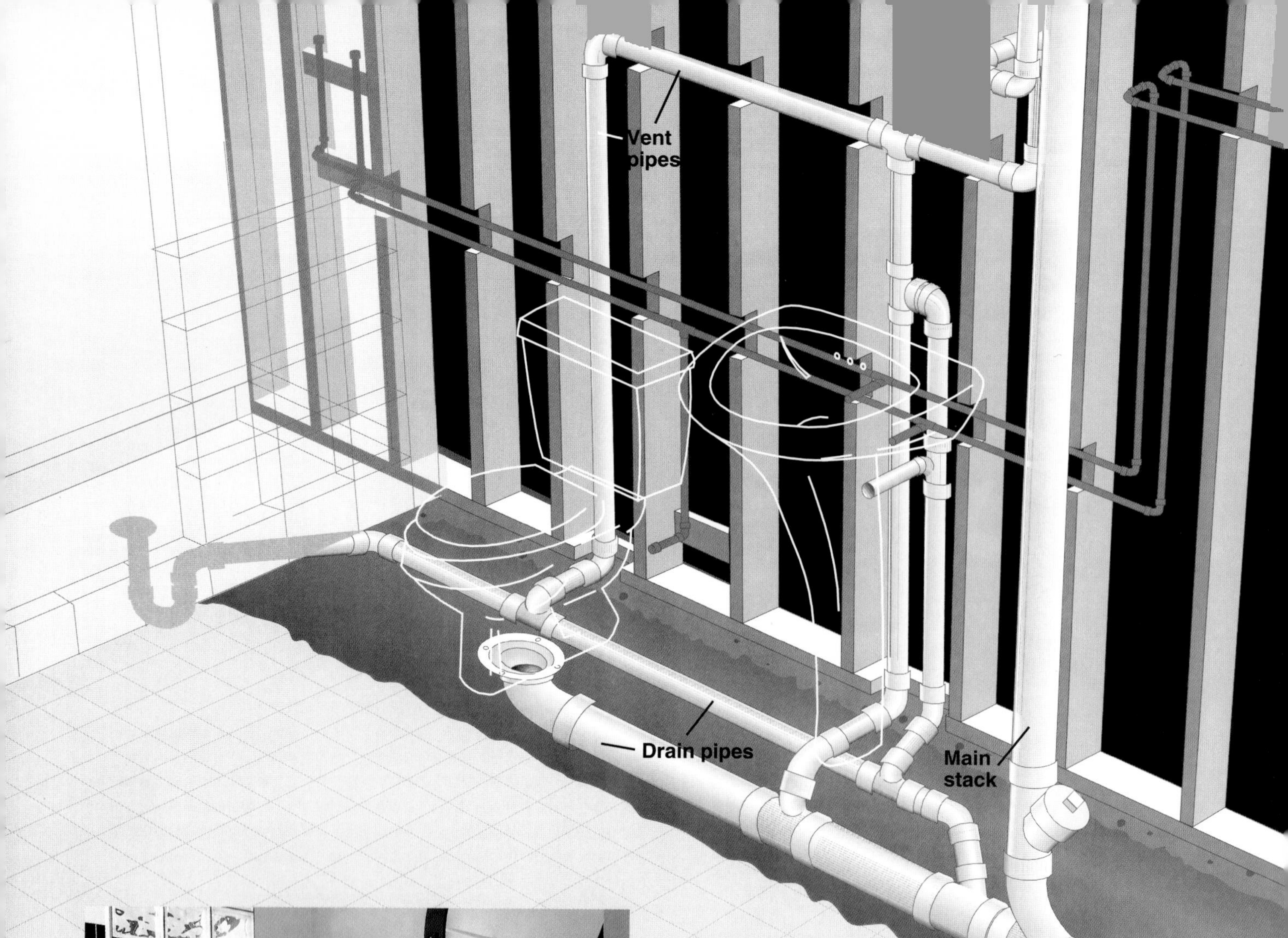

Our demonstration bathroom includes a shower, toilet, and vanity sink arranged in a line to simplify trenching. A 2" drain pipe services the new shower and sink; a 3" pipe services the new toilet. The drain pipes converge at a Y-fitting joined to the existing main drain. The shower, toilet, and sink have individual vent pipes that meet inside the wet wall before extending up into the attic, where they join the main waste-vent stack.

Plumbing a Basement Bath

When installing a basement bath, make sure you allow extra time for tearing out the concrete floor to accommodate the drains, and for construction of a wet wall to enclose supply and vent pipes. Constructing your wet wall with 2 × 6 studs and plates will provide ample room for running pipes. Be sure to schedule an inspection by a building official before you replace the concrete and cover the walls with wallboard.

Whenever possible, try to hold down costs by locating your basement bath close to existing drains and supply pipes.

How to Plumb a Basement Bath

1 Outline a 24"-wide trench on the concrete where new branch drains will run to the main drain. In our project, we ran the trench parallel to an outside wall, leaving a 6" ledge for framing a wet wall. Use a masonry chisel and hand maul to break up concrete near the stack.

2 Use a circular saw and masonry blade to cut along the outline, then break the rest of the trench into convenient chunks with a jackhammer. Remove any remaining concrete with a chisel. Excavate the trench to a depth about 2" deeper than the main drain. At vent locations for the shower and toilet, cut 3" notches in the concrete all the way to the wall.

3 Cut the 2 × 6 framing for the wet wall that will hold the pipes. Cut 3" notches in the bottom plate for the pipes, then secure the plate to the floor with construction adhesive and masonry nails. Install the top plate, then attach studs.

4 Assemble a 2" horizontal drain pipe for the sink and shower, and a 3" drain pipe for the toilet. The 2" drain pipe includes a solvent-glued trap for the shower, a vent T, and a waste T for the sink drain. The toilet drain includes a toilet bend and a vent T. Use elbows and straight lengths of pipe to extend the vent and drain pipes to the wet wall. Make sure the vent fittings angle upward from the drain pipe at least 45°.

(continued next page)

How to Plumb a Basement Bath (continued)

5 Use pairs of stakes with vinyl support straps slung between them to cradle drain pipes in the proper position (inset). The drain pipes should be positioned so they slope ¼" per foot down toward the main drain.

6 Assemble the fittings required to tie the new branch drains into the main drain. In our project, we will be cutting out the cleanout and sweep on the main waste-vent stack in order to install a new assembly that includes a Y-fitting to accept the two new drain pipes.

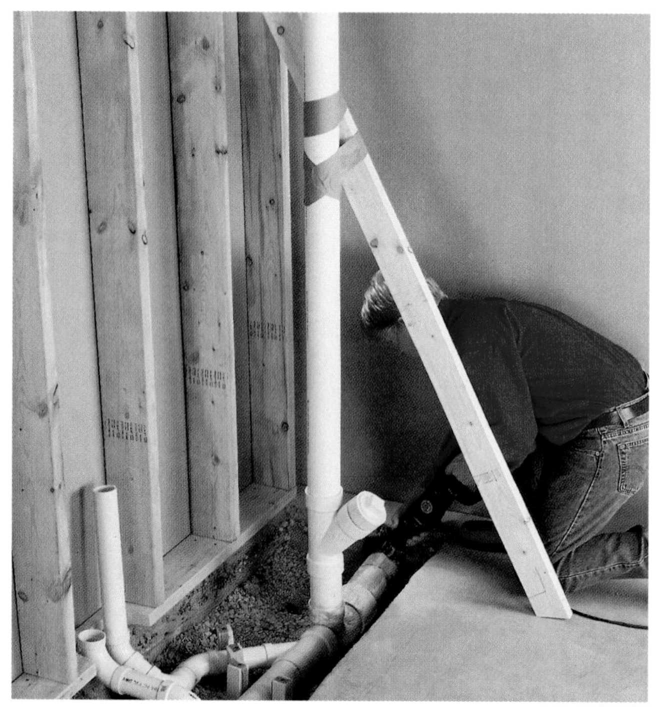

7 Support the main waste-vent stack before cutting. Use a 2 × 4 for a plastic stack, or riser clamps (page 97) for a cast-iron stack. Using a reciprocating saw (or cast iron cutter), cut into the main drain as close as possible to the stack.

8 Cut into the stack above the cleanout and remove the pipe and fittings. Wear rubber gloves, and have a bucket and plastic bags ready, as old pipes and fittings may be coated with messy sludge.

9 Test-fit, then solvent-glue the new cleanout and reducing Y assembly into the main drain. Support the weight of the stack by adding sand underneath the Y, but leave plenty of space around the end for connecting the new branch pipes.

10 Working from the reducing Y, solvent-glue the new drain pipes together. Be careful to maintain proper slope of the drain pipes when gluing. Be sure the toilet and shower drains extend at least 2" above the floor level.

11 Check for leaks by pouring fresh water into each new drain pipe. If no leaks appear, cap or plug the drains with rags to prevent sewer gas from leaking into the work area as you complete the installation.

(continued next page)

12 Run 2" vent pipes from the drains up the inside of the wet wall. Notch the studs and insert a horizontal vent pipe, then attach the vertical vent pipes with an elbow and vent T-fitting. Test-fit all pipes, then solvent-glue them in place.

13 Route the vent pipe from the wet wall to a point below a wall cavity running from the basement to the attic. NOTE: If there is an existing vent pipe in the basement, you can tie into this pipe rather than run the vent to the attic.

14 If you are running vent pipes in a two-story home, remove sections of wall surface as needed to bore holes for running the vent pipe through wall plates. Feed the vent pipe up into the wall cavity from the basement.

15 Wedge the vent pipe in place while you solvent-glue the fittings. Support the vent pipe at each floor with vinyl pipe straps. Do not patch the walls until your work has been inspected by a building official.

16 Cut into the main stack in the attic, and install a vent T-fitting, using banded couplings. (If the stack is cast iron, make sure to support it adequately above and below the cuts.) Attach a test T-fitting to the vent T, then join the new vent pipe to the stack, using elbows and lengths of straight pipe as needed.

17 Shut off the main water supply, cut into the water supply pipes as near as possible to the new bathroom, and install T-fittings. Install full-bore control valves on each line, then run ¾" branch supply pipes down into the wet wall by notching the top wall plate. Extend the pipes across the wall by notching the studs.

18 Use reducing T-fittings to run ½" supplies to each fixture, ending with female threaded adapters. Install backing boards, and strap the pipes in place. Attach metal protector plates over notched studs to protect pipes. After having your work approved by a building official, fill in around the pipes with dirt or sand, then mix and pour new concrete to cover the trench. Trowel the surface smooth, and let the cement cure for 3 days before installing fixtures.

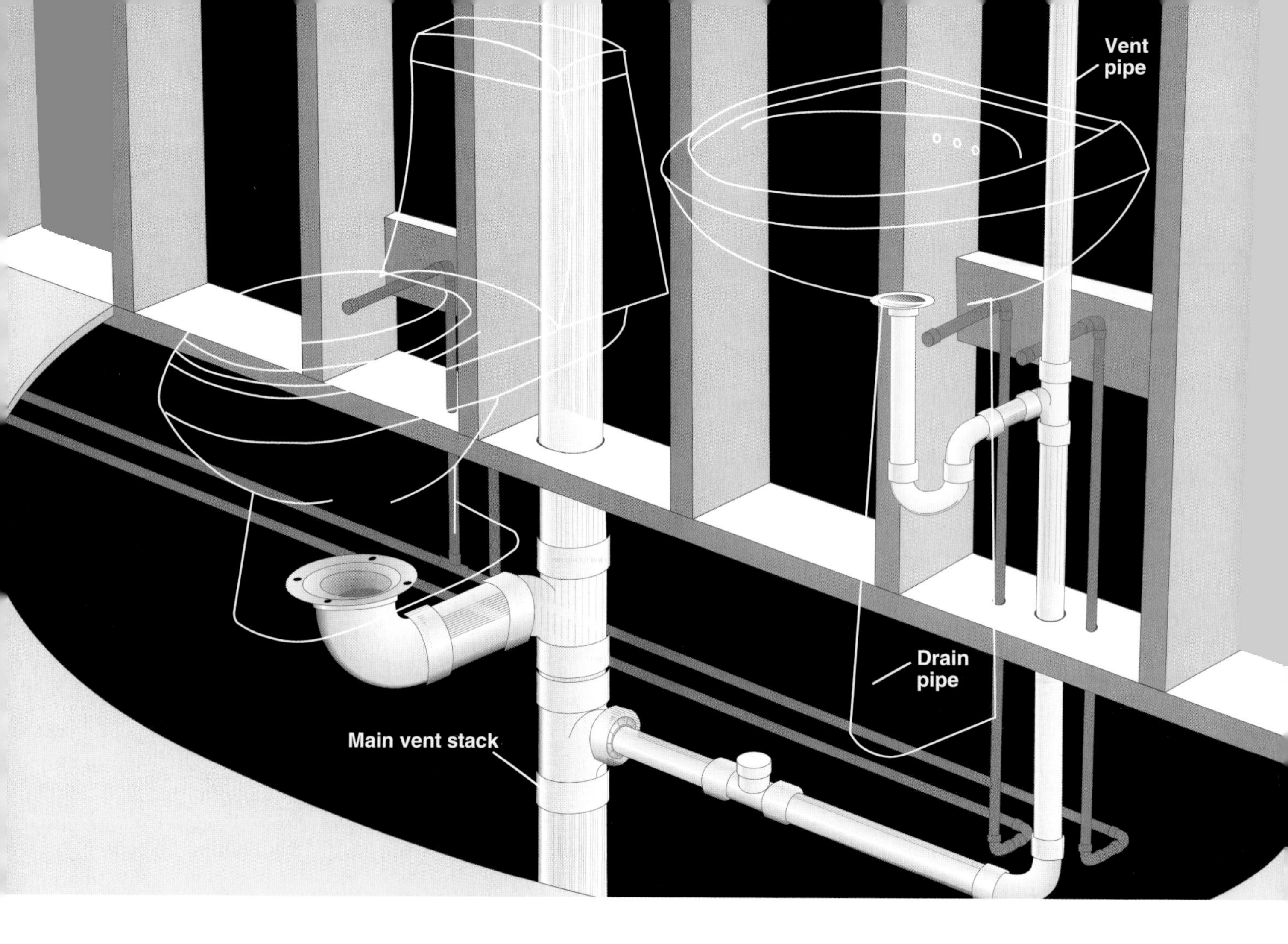

Plumbing a Half Bath

A first-story half bath is easy to install when located behind a kitchen or existing bathroom, because you can take advantage of accessible supply and DWV lines. It is possible to add a half bath on an upper story or in a location distant from existing plumbing, but the complexity and cost of the project may be increased considerably.

Be sure that the new fixtures are adequately vented. We vented the pedestal sink with a pipe that runs up the wall a few feet before turning to join the main stack. However, if there are higher fixtures draining into the main stack, you would be required to run the vent up to a point at least 6" above the highest fixture before splicing it into the main stack or an existing vent pipe. When the toilet is located within 6 ft. of the stack, as in our design, it requires no additional vent pipe.

The techniques for plumbing a half bath are similar to those used for a master bathroom. Refer to pages 50 to 57 for more detailed information.

In our half bath, the toilet and sink are close to the main stack for ease of installation, but are spaced far enough apart to meet minimum allowed distances between fixtures. Check your local Code for any restrictions in your area. Generally, there should be at least 15" from the center of the toilet drain to a side wall or fixture, and a minimum of 21" of space between the front edge of the toilet and the wall.

How to Plumb a Half Bath

1 Locate the main waste-vent stack in the wet wall, and remove the wall surface behind the planned location for the toilet and sink. Cut a 4½"-diameter hole for the toilet flange (centered 12" from the wall, for most toilets). Drill two ¾" holes through the sole plate for sink supply lines and one hole for the toilet supply line. Drill a 2" hole for the sink drain.

2 In the basement, cut away a section of the stack and insert two waste T-fittings. The top fitting should have a 3" side inlet for the toilet drain; the bottom fitting requires a 1½" reducing bushing for the sink drain. Install a toilet bend and 3" drain pipe for the toilet, and install a 1½" drain pipe with a sweep elbow for the sink.

3 Tap into water distribution pipes with ¾" × ½" reducing T-fittings, then run ½" copper supply pipes through the holes in the sole plate to the sink and toilet. Support all pipes at 4-ft. intervals with strapping attached to joists.

4 Attach drop ear elbows to the ends of the supply pipes, and anchor them to blocking installed between studs. Anchor the drain pipe to the blocking, then run a vertical vent pipe from the waste T-fitting up the wall to a point at least 6" above the highest fixture on the main stack. Then, route the vent pipe horizontally and splice it into the vent stack with a vent T.

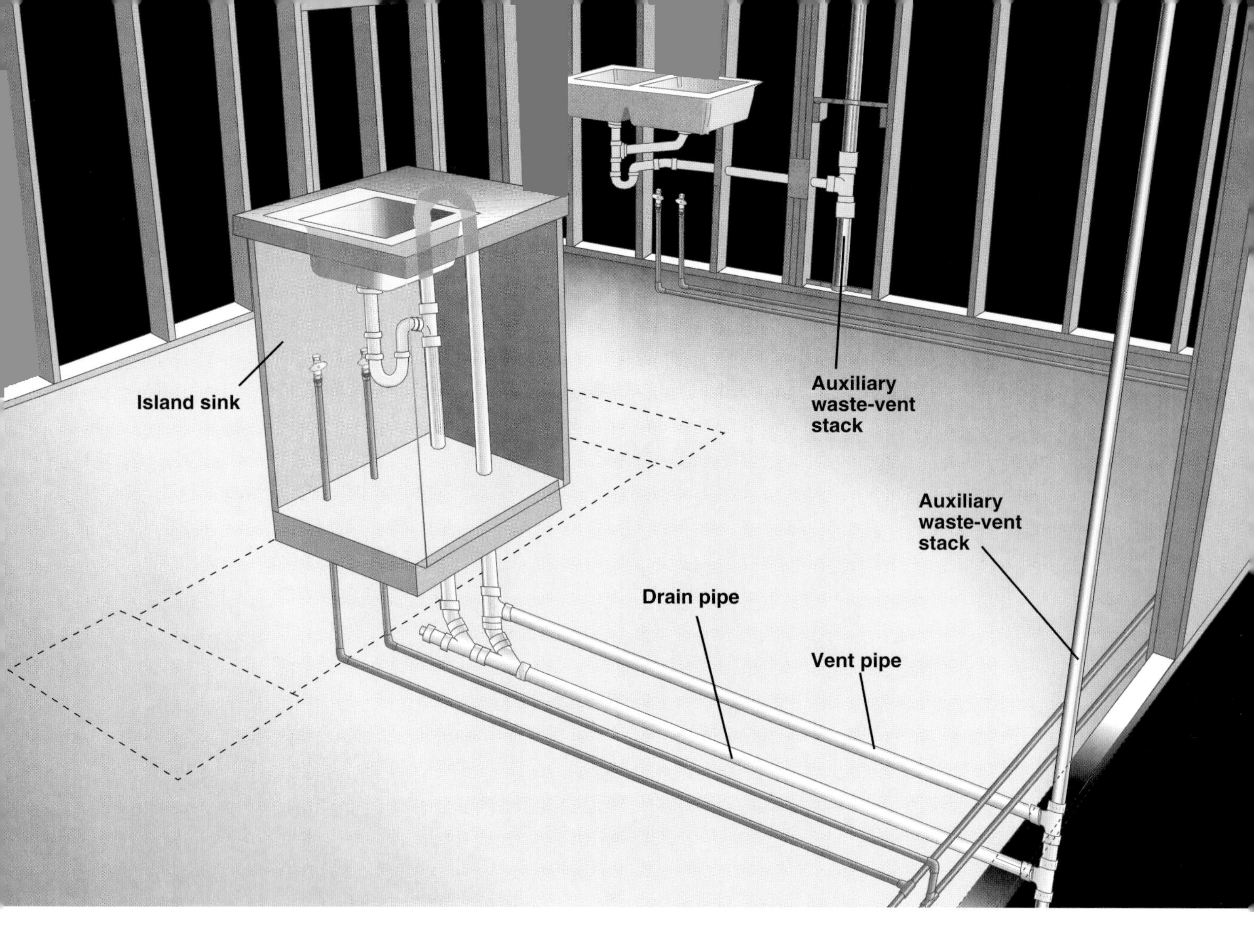

Plumbing a Kitchen

Plumbing a remodeled kitchen is a relatively easy job if your kitchen includes only a wall sink. If your project includes an island sink, however, the work becomes more complicated.

An island sink poses problems because there is no adjacent wall in which to run a vent pipe. For an island sink, you will need to use a special plumbing configuration known as a *loop vent*.

Each loop vent situation is different, and your configuration will depend on the location of existing waste-vent stacks, the direction of the floor joists, and the size and location of your sink base cabinet. Consult your local plumbing inspector for help in laying out the loop vent.

For our demonstration kitchen, we have divided the project into three phases:

- How to Install DWV Pipes for a Wall Sink (pages 68 to 70)
- How to Install DWV Pipes for an Island Sink (pages 71 to 75)
- How to Install New Supply Pipes (pages 76 to 77)

Our demonstration kitchen includes a double wall sink and an island sink. The 1½" drain for the wall sink connects to an existing 2" galvanized waste-vent stack; since the trap is within 3½ ft. of the stack, no vent pipe is required. The drain for the island sink uses a loop vent configuration connected to an auxiliary waste-vent stack in the basement.

Tips for Plumbing a Kitchen

Insulate exterior walls if you live in a region with freezing winter temperatures. Where possible, run water supply pipes through the floor or interior partition walls, rather than exterior walls.

Use existing waste-vent stacks to connect the new DWV pipes. In addition to a main waste-vent stack, most homes have one or more auxiliary waste-vent stacks in the kitchen that can be used to connect new DWV pipes.

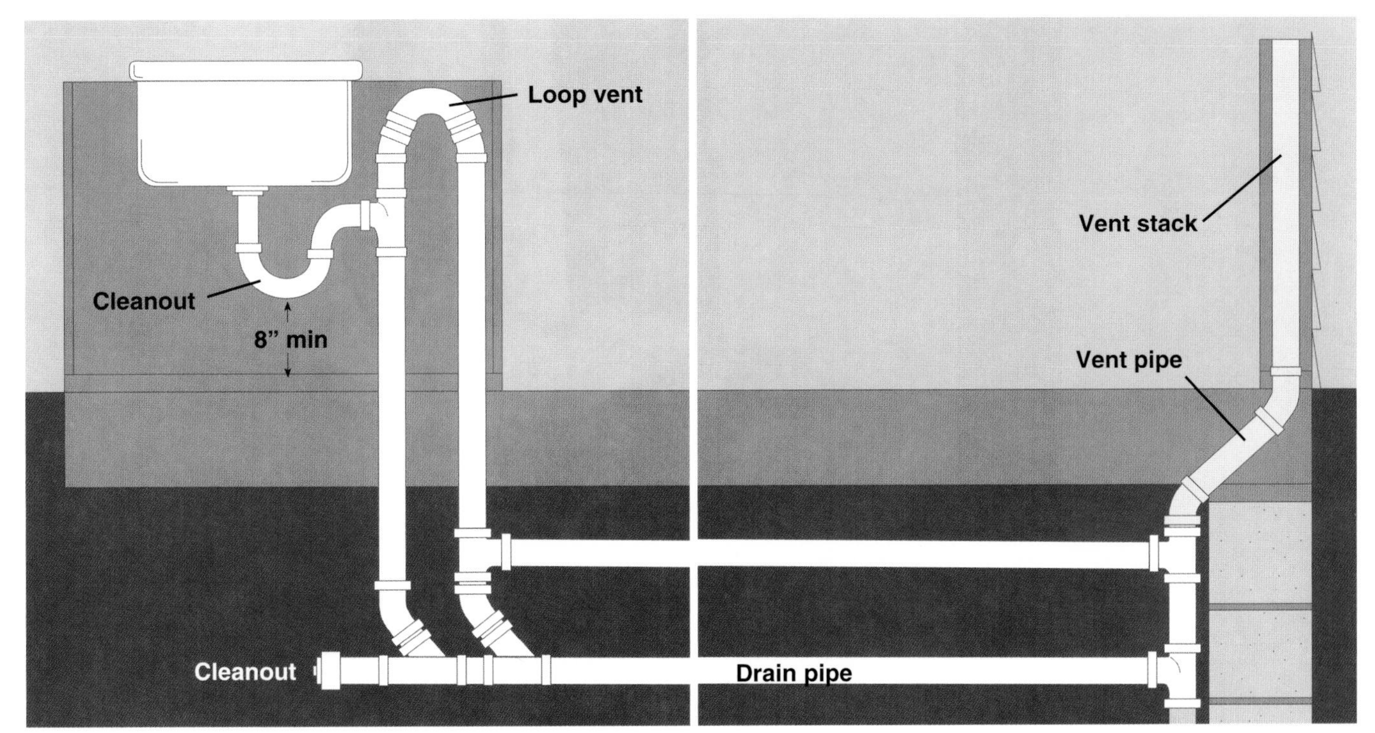

Loop vent makes it possible to vent a sink when there is no adjacent wall to house the vent pipe. The drain is vented with a loop of pipe that arches up against the countertop and away from the drain before dropping through the floor. The vent pipe then runs horizontally to an existing vent pipe. In our project, we have tied the island vent to a vent pipe extending up from a basement utility sink. NOTE: Loop vents are subject to local Code restrictions. Always consult your building inspector for guidelines on venting an island sink.

How to Install DWV Pipes for a Wall Sink

1 Determine the location of the sink drain by marking the position of the sink and base cabinet on the floor. Mark a point on the floor indicating the position of the sink drain opening. This point will serve as a reference for aligning the sink drain stub-out.

2 Mark a route for the new drain pipe through the studs behind the wall sink cabinet. The drain pipe should angle ¼" per foot down toward the waste-vent stack.

3 Use a right-angle drill and hole saw to bore holes for the drain pipe (page 46). On non-loadbearing studs, such as the cripple studs beneath a window, you can notch the studs with a reciprocating saw to simplify the installation of the drain pipe. If the studs are loadbearing, however, you must thread the run though the bored holes, using couplings to join short lengths of pipe as you create the run.

4 Measure, cut, and dry-fit a horizontal drain pipe to run from the waste-vent stack to the sink drain stub-out. Create the stub-out with a 45° elbow and 6" length of 1½" pipe. NOTE: If the sink trap in your installation will be more than 3½ ft. from the waste-vent pipe, you will need to install a waste T and run a vent pipe up the wall, connecting it to the vent stack at a point at least 6" above the lip of the sink.

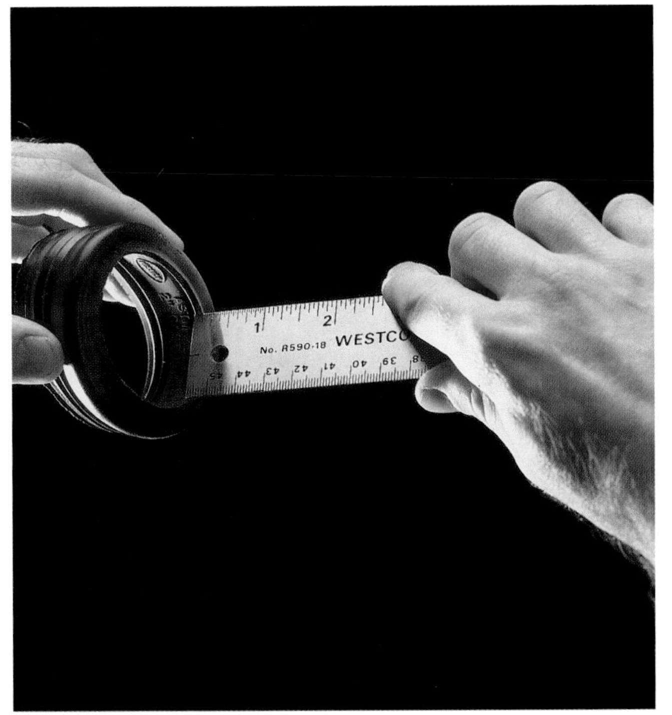

5 Remove the neoprene sleeve from a banded coupling, then roll the lip back and measure the thickness of the separator ring.

6 Attach two lengths of 2" pipe, at least 4" long, to the top and bottom openings on a 2" × 2" × 1½" waste T. Hold the fitting alongside the waste-vent stack, then mark the stack for cutting, allowing space for the separator rings on the banded couplings.

(continued next page)

How to Install DWV Pipes for a Wall Sink (continued)

7 Use riser clamps and 2 × 4 blocking to support the waste-vent stack above and below the new drain pipe, then cut out the waste-vent stack along the marked lines, using a reciprocating saw and metal-cutting blade.

8 Slide banded couplings onto the cut ends of the waste-vent stack, and roll back the lips of the neoprene sleeves. Position the waste T assembly, then roll the sleeves into place over the plastic pipes.

9 Slide the metal bands into place over the neoprene sleeves, and tighten the clamps with a ratchet wrench or screwdriver.

10 Solvent-glue the drain pipe, beginning at the waste-vent stack. Use a 90° elbow and a short length of pipe to create a drain stub-out extending about 4" out from the wall.

How to Install DWV Pipes for an Island Sink

1 Position the base cabinet for the island sink, according to your kitchen plans. Mark the cabinet position on the floor with tape, then move the cabinet out of the way.

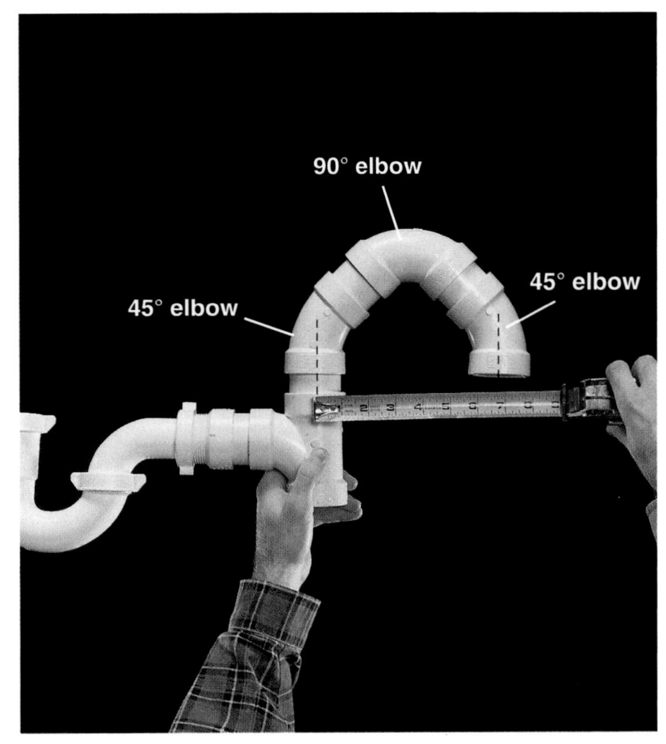

2 Create the beginning of the drain and loop vent by test-fitting a drain trap, waste T, two 45° elbows, and a 90° elbow, linking them with 2" lengths of pipe. Measure the width of the loop between the centerpoints of the fittings.

3 Draw a reference line perpendicular to the wall to use as a guide when positioning the drain pipes. A cardboard template of the sink can help you position the loop vent inside the outline of the cabinet.

4 Position the loop assembly on the floor, and use it as a guide for marking hole locations. Make sure to position the vent loop so the holes are not over joists.

(continued next page)

How to Install DWV Pipes for an Island Sink (continued)

5 Use a hole saw with a diameter slightly larger than the vent pipes to bore holes in the subfloor at the marked locations. Note the positions of the holes by carefully measuring from the edges of the taped cabinet outline; these measurements will make it easier to position matching holes in the floor of the base cabinet.

6 Reposition the base cabinet, and mark the floor of the cabinet where the drain and vent pipes will run. (Make sure to allow for the thickness of the cabinet sides when measuring.) Use the hole saw to bore holes in the floor of the cabinet, directly above the holes in the subfloor.

7 Measure, cut, and assemble the drain and loop vent assembly. Tape the top of the loop in place against a brace laid across the top of the cabinet, then extend the drain and vent pipes through the holes in the floor of the cabinet. The waste T should be about 18" above the floor, and the drain and vent pipes should extend about 2 ft. through the floor.

8 In the basement, establish a route from the island vent pipe to an existing vent pipe. (In our project, we are using the auxiliary waste-vent stack near a utility sink.) Hold a long length of pipe between the pipes, and mark for T-fittings. Cut off the plastic vent pipe at the mark, then dry-fit a waste T-fitting to the end of the pipe.

9 Hold a waste T against the vent stack, and mark the horizontal vent pipe at the correct length. Fit the horizontal pipe into the waste T, then tape the assembly in place against the vent stack. The vent pipe should angle ¼" per foot down toward the drain.

10 Fit a 3" length of pipe in the bottom opening on the T-fitting attached to the vent pipe, then mark both the vent pipe and the drain pipe for 45° elbows. Cut off the drain and vent pipes at the marks, then dry-fit the elbows onto the pipes.

11 Extend both the vent pipe and drain pipe by dry-fitting 3" lengths of pipe and Y-fittings to the elbows. Using a carpenter's level, make sure the horizontal drain pipe will slope toward the waste-vent at a pitch of ¼" per ft. Measure and cut a short length of pipe to fit between the Y-fittings.

(continued next page)

How to Install DWV Pipes for an Island Sink (continued)

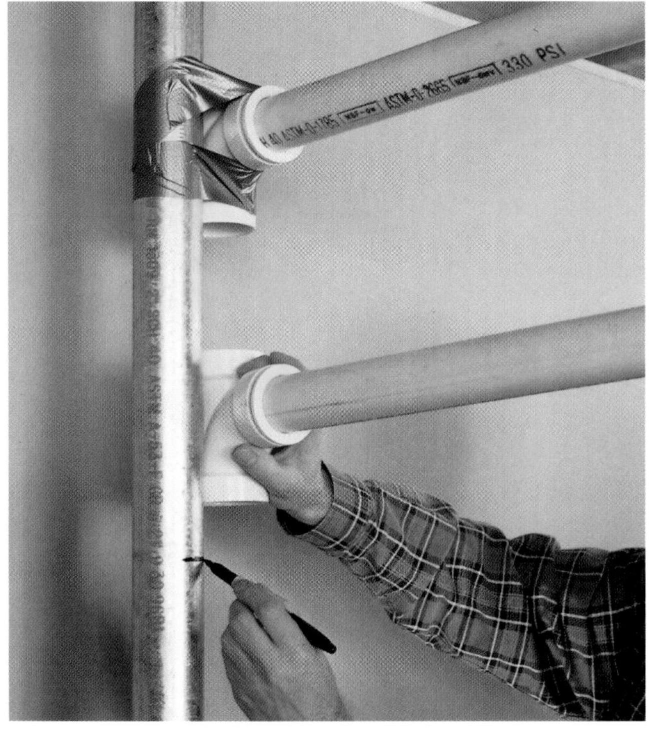

12 Cut a horizontal drain pipe to reach from the vent Y-fitting to the auxiliary waste-vent stack. Attach a waste T to the end of the drain pipe, then position it against the drain stack, maintaining a downward slope of ¼" per ft. Mark the auxiliary stack for cutting above and below the fittings.

13 Cut out the auxiliary stack at the marks. Use the T-fittings and short lengths of pipe to assemble an insert piece to fit between the cutoff ends of the auxiliary stack. The insert assembly should be about ½" shorter than the removed section of stack.

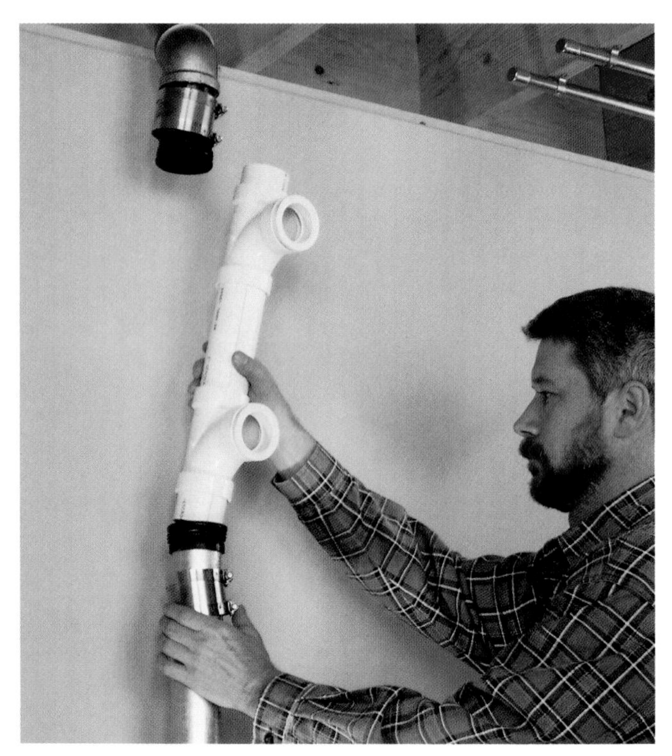

14 Slide banded couplings onto the cut ends of the auxiliary stack, then insert the plastic pipe assembly and loosely tighten the clamps.

15 At the open inlet on the drain pipe Y-fitting, insert a cleanout fitting.

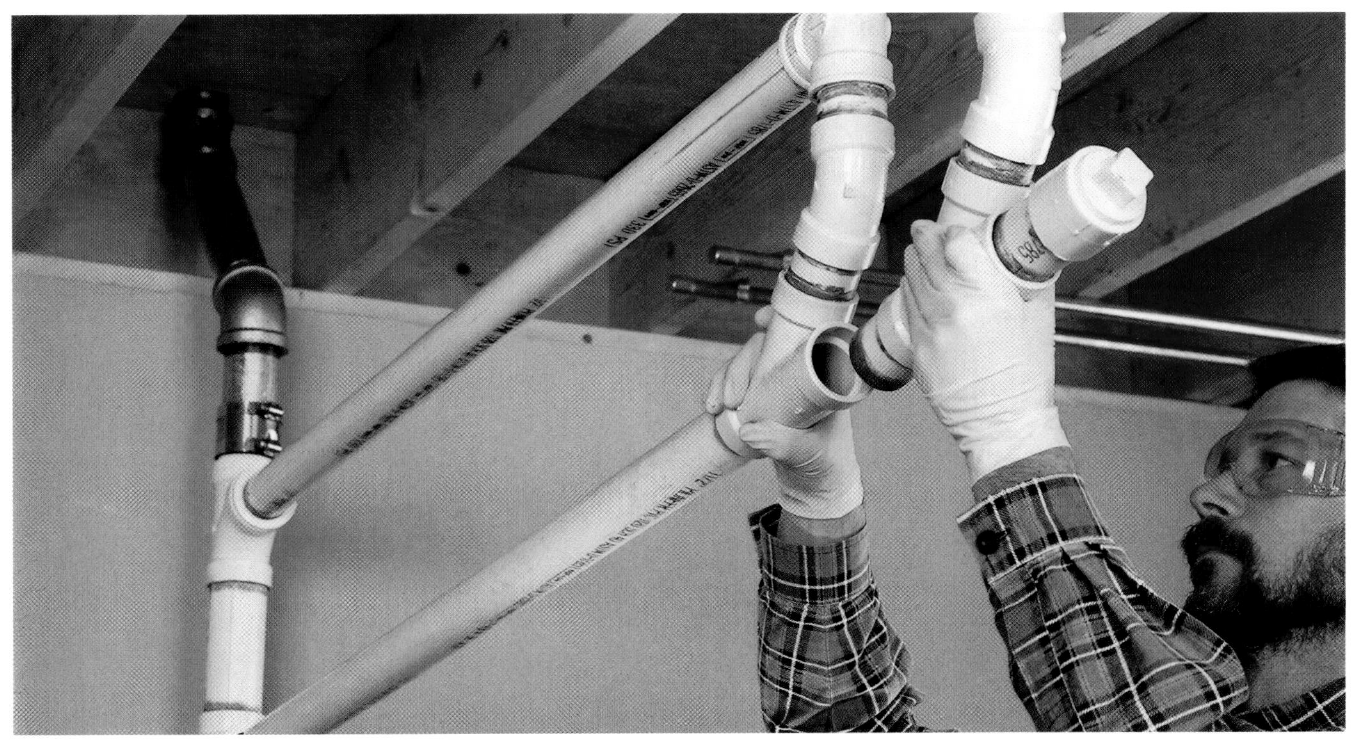

16 Solvent-glue all pipes and fittings found in the basement, beginning with the assembly inserted into the existing waste-vent stack, but do not glue the vertical drain and vent pipes running up into the cabinet. Tighten the banded couplings at the auxiliary stack. Support the horizontal pipes every 4 ft. with strapping nailed to the joists, then detach the vertical pipes extending up into the island cabinet. The final connection for the drain and vent loop will be completed as other phases of the kitchen remodeling project are finished.

17 After installing flooring and attaching cleats for the island base cabinet, cut away the flooring covering the holes for the drain and vent pipes.

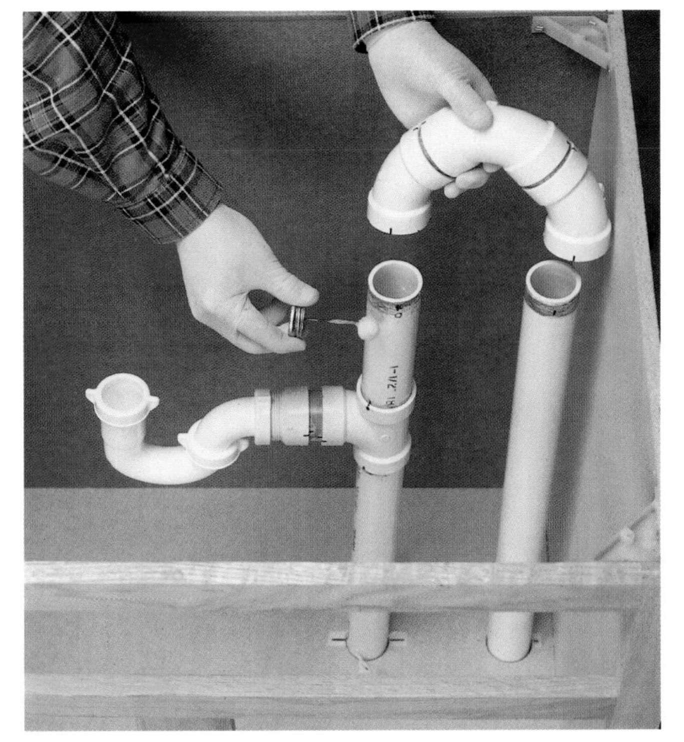

18 Install the base cabinet, then insert the drain and vent pipes through the holes in the cabinet floor and solvent-glue the pieces together.

How to Install New Supply Pipes

1 Drill two 1"-diameter holes, spaced about 6" apart, through the floor of the island base cabinet and the underlying subfloor. Position the holes so they are not over floor joists. Drill similar holes in the floor of the base cabinet for the wall sink.

2 Turn off the water at the main shutoff, and drain the pipes. Cut out any old water supply pipes that obstruct new pipe runs, using a tubing cutter or hacksaw. In our project, we are removing the old pipe back to a point where it is convenient to begin the new branch lines.

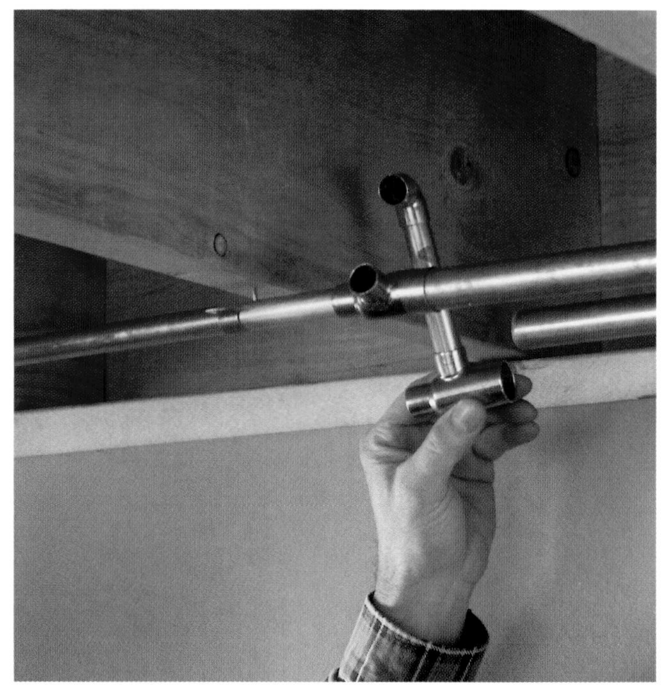

3 Dry-fit T-fittings on each supply pipe (we used ¾" × ½" × ½" reducing T-fittings). Use elbows and lengths of copper pipe to begin the new branch lines running to the island sink and the wall sink. The parallel pipes should be routed so they are between 3" and 6" apart.

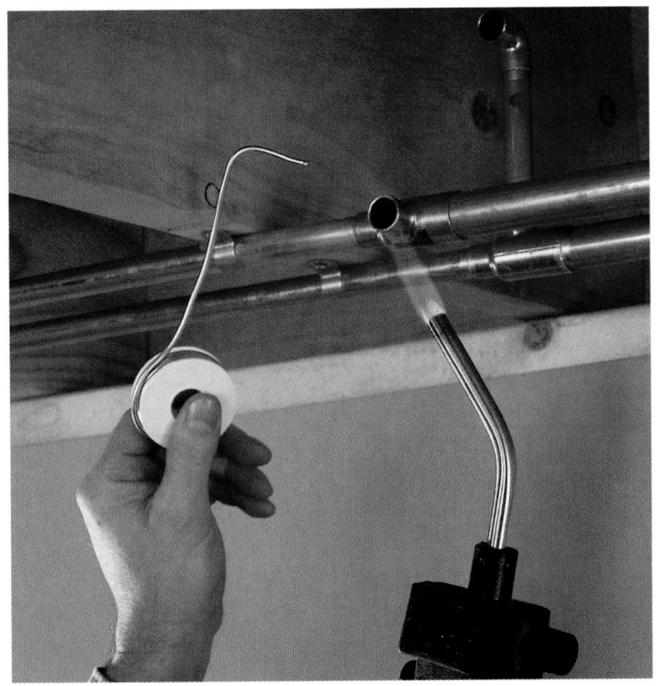

4 Solder the pipes and fittings together, beginning at the T-fittings. Support the horizontal pipe runs every 6 ft. with strapping attached to joists.

5 Extend the branch lines to points directly below the holes leading up into the base cabinets. Use elbows and lengths of pipe to form vertical risers extending at least 12" into the base cabinets. Use a small level to position the risers so they are plumb, then mark the pipe for cutting.

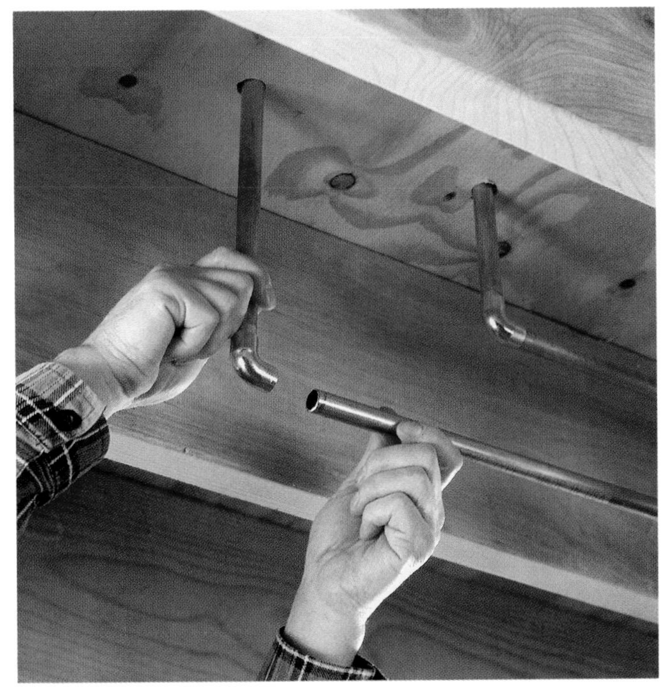

6 Fit the horizontal pipes and risers together, and solder them in place. Install blocking between joists, and anchor the risers to the blocking with pipe straps.

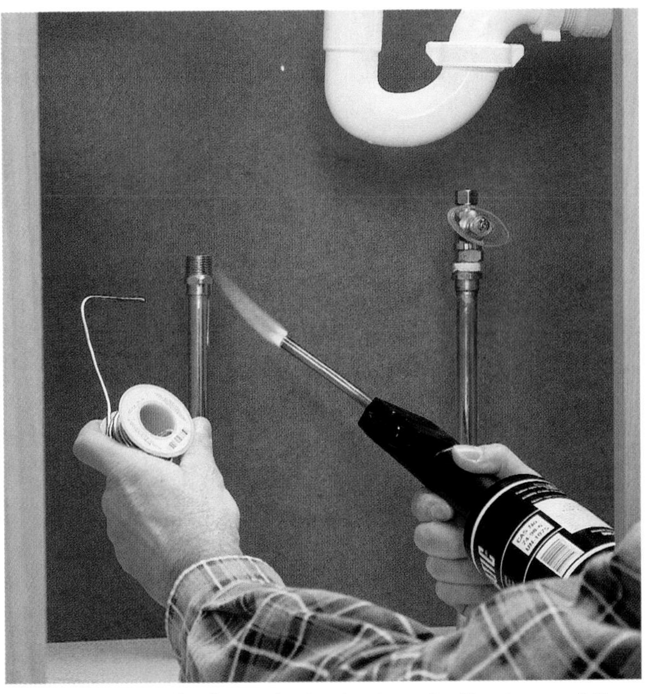

7 Solder male threaded adapters to the tops of the risers, then screw threaded shutoff valves onto the fittings.

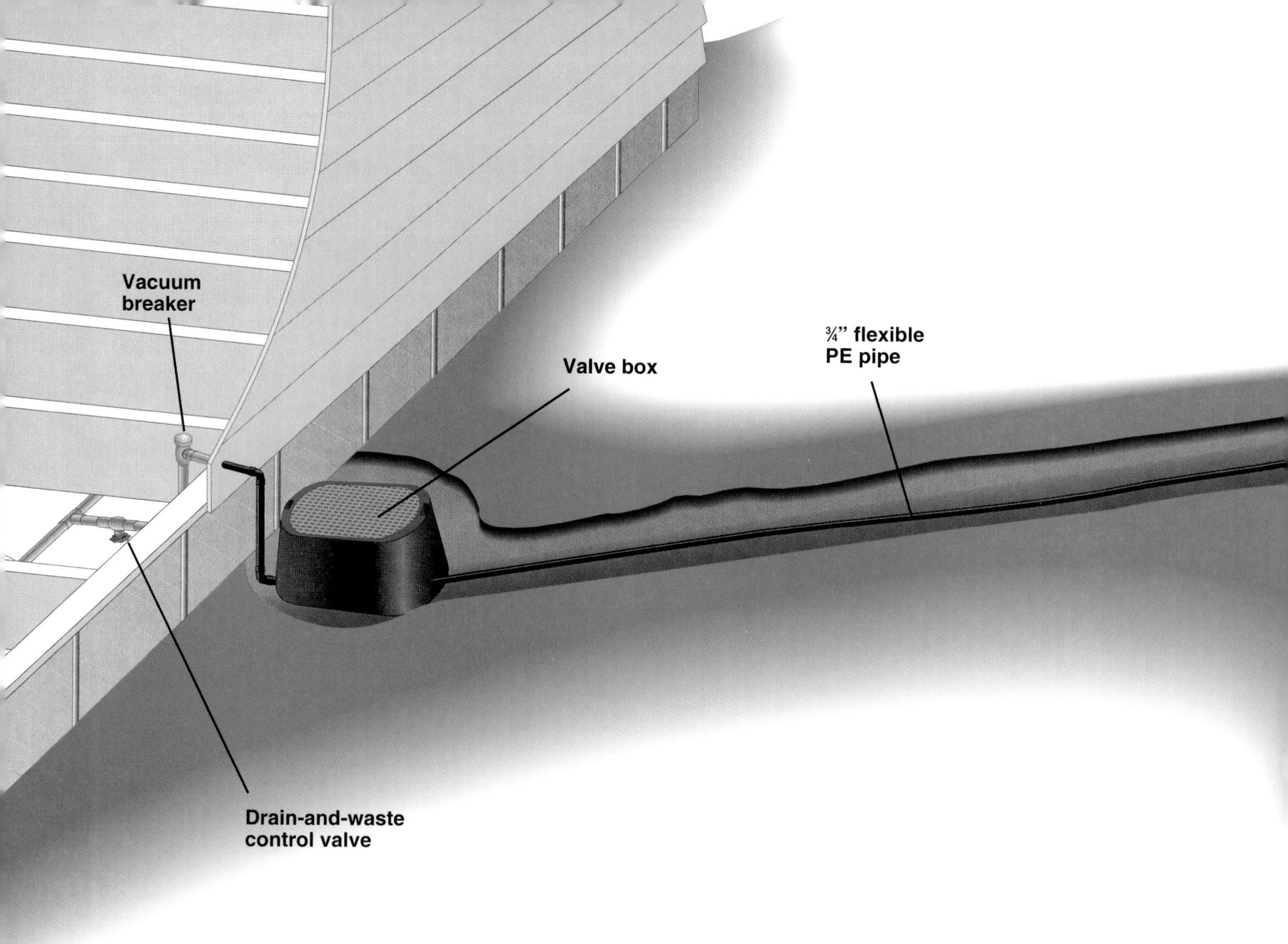

Installing Outdoor Plumbing

Flexible polyethylene (PE) pipe (page 30) is used to extend cold-water plumbing lines to outdoor fixtures, such as a sink located in a shed or detached garage, a lawn sprinkler system, or garden spigot. In mild climates, outdoor plumbing can remain in service year-round, but in regions with a frost season, the outdoor supply pipes must be drained or blown empty with pressurized air to prevent the pipes from rupturing when the ground freezes in winter.

On the following pages you will see how to run supply pipes from the house to a utility sink in a detached garage. The utility sink drains into a rock-filled *dry well* installed in the yard. A dry well is designed to handle only "gray water" waste, such as the soapy rinse water created by washing tools or work clothes. Never use a dry well drain for septic materials, such as animal waste or food scraps. Never pour paints, solvent-based liquids, or solid materials into a sink that drains into a dry well. Such materials will quickly clog up your system and will eventually filter down into the groundwater supply.

Like an indoor sink, the garage utility sink has a vent pipe running up from the drain trap. This vent can extend through the roof (page 101), or it can be extended through the side wall of the garage and covered with a screen to keep birds and insects out.

Before digging a trench for an outdoor plumbing line, contact your local utility companies and ask them to mark the locations of underground gas, power, telephone, and water lines.

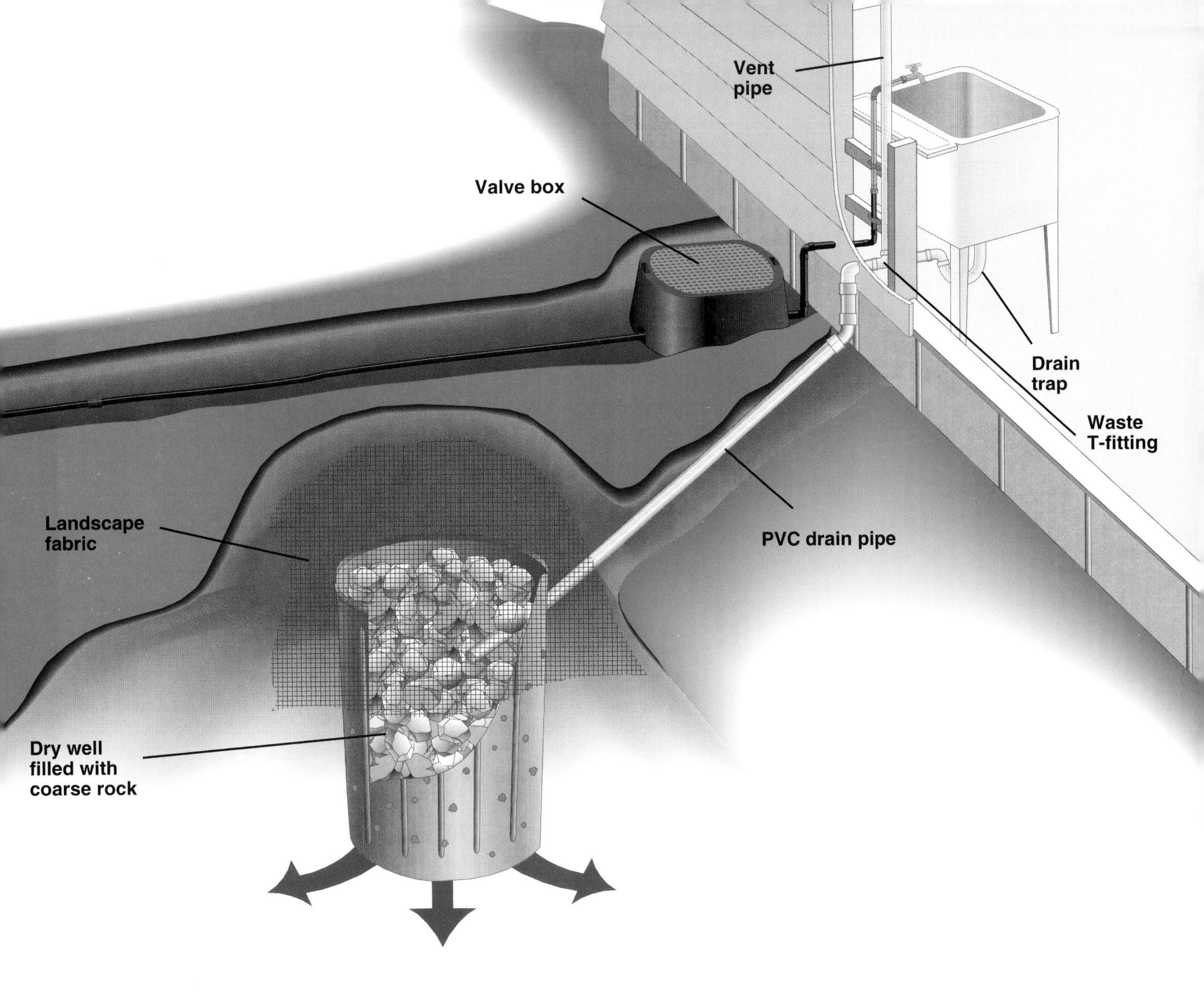

Underground lawn sprinkler systems can be installed using the same basic techniques used for plumbing an outdoor utility sink. Sprinkler systems vary from manufacturer to manufacturer, so make sure to follow the product recommendations when planning your system. See page 83.

How to Install Outdoor Plumbing for a Garage Sink

1 Plan a convenient route from a basement ¾" cold-water supply pipe to the outdoor sink location, then drill a 1"-diameter hole through the sill plate. Drill a similar hole where the pipe will enter the garage. On the ground outside, lay out the pipe run with spray paint or stakes.

2 Use a flat spade to remove sod for an 8"- to 12"-wide trench along the marked route from the house to the garage. Set the sod aside and keep it moist so it can be reused after the project is complete. Dig a trench that slopes slightly (⅛" per foot) toward the house and is at least 10" deep at its shallowest point. Use a long straight 2 × 4 and level to ensure that the trench has the correct slope.

3 Below the access hole in the rim joist, dig a small pit and install a plastic valve box so the top is flush with the ground. Lay a thick layer of gravel in the bottom of the box. Dig a similar pit and install a second valve box at the opposite end of the trench, where the water line will enter the garage.

4 Run ¾" PE pipe along the bottom of the trench from the house to the utility sink location. Use insert couplings and stainless steel clamps when it is necessary to join two lengths of pipe.

TIP: To run pipe under a sidewalk, attach a length of rigid PVC pipe to a garden hose with a pipe-to-hose adapter. Cap the end of the pipe, and drill a ⅛" hole in the center of the cap. Turn on the water, and use the high-pressure stream to bore a tunnel.

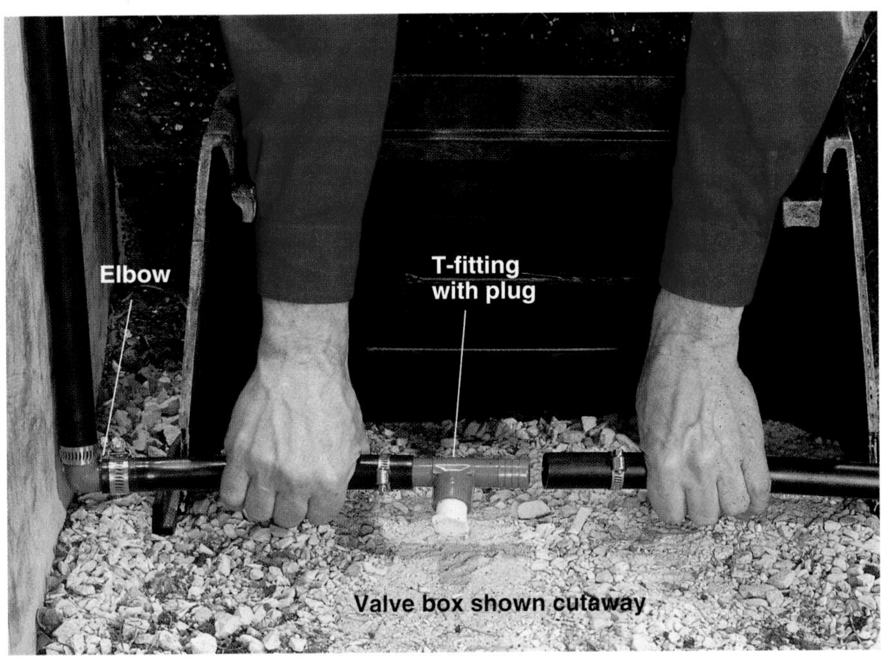

5 At each end of the trench, extend the pipe through the valve box and up the foundation wall, using a barbed elbow fitting to make the 90° bend. Install a barbed T-fitting with a threaded outlet in the valve box, so the threaded portion of the fitting faces down. Insert a male threaded plug in the bottom outlet of the T-fitting.

6 Use barbed elbow fittings to extend the pipe into the basement and garage, then use pipe straps and masonry screws to anchor the PE pipe to the foundation.

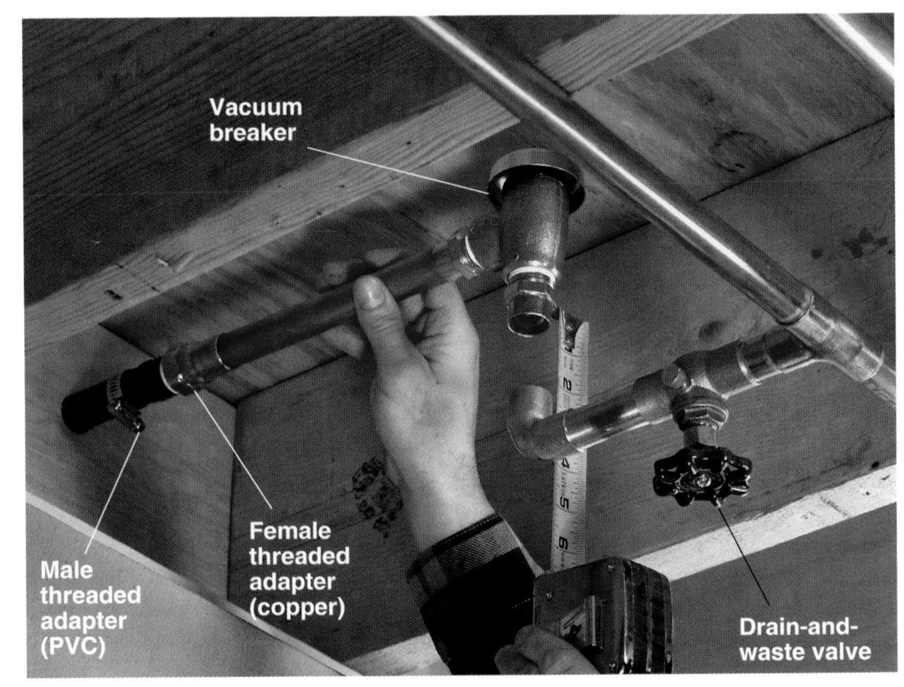

7 Inside the house, make the transition between the PE pipe and the copper cold-water supply pipe, using a threaded male PVC adapter, a female threaded copper adapter, a vacuum breaker, a drain-and-waste valve, and a copper T-fitting, as shown. The drain-and-waste valve includes a threaded cap, which can be removed to blow water from the lines when you are winterizing the system.

(continued next page)

How to Install Outdoor Plumbing for a Garage Sink (continued)

8 In the garage, attach a male threaded PVC adapter to the end of the PE pipe, then use a copper female threaded adapter, elbow, and male threaded adapter to extend a copper riser up to a brass hose bib. After completing the supply pipe installation, fill in the trench, tamping the soil firmly. Install the utility sink, complete with 1½" drain trap and waste T-fitting (page 79). Bore a 2" hole in the wall where the sink drain will exit the garage.

9 At least 6 ft. from the garage, dig a pit about 2 ft. in diameter and 3 ft. deep. Punch holes in the sides and bottom of an old trash can, and cut a 2" hole in the side of the can, about 4" from the top edge. Insert the can into the pit; the top edge should be about 6" below ground level. Run 1½" PVC drain pipe from the utility sink to the dry well (page 28). Fill the dry well with coarse rock, drape landscape fabric over it, then cover the trench and well with soil and reinstall the sod. Extend a vent pipe up from the waste T through the roof or side wall of the garage (page 101).

How to Winterize Outdoor Plumbing Pipes in Cold Climates

Close the drain-and-waste valve for the outdoor supply pipe, then remove the cap on the drain nipple. With the hose bib on the outdoor sink open, attach an air compressor to the valve nipple, then blow water from the system using no more than 50 psi (pounds per square inch) of air pressure. Remove the plugs from the T-fittings in each valve box, and store them for the winter.

Components of an Underground Sprinkler System

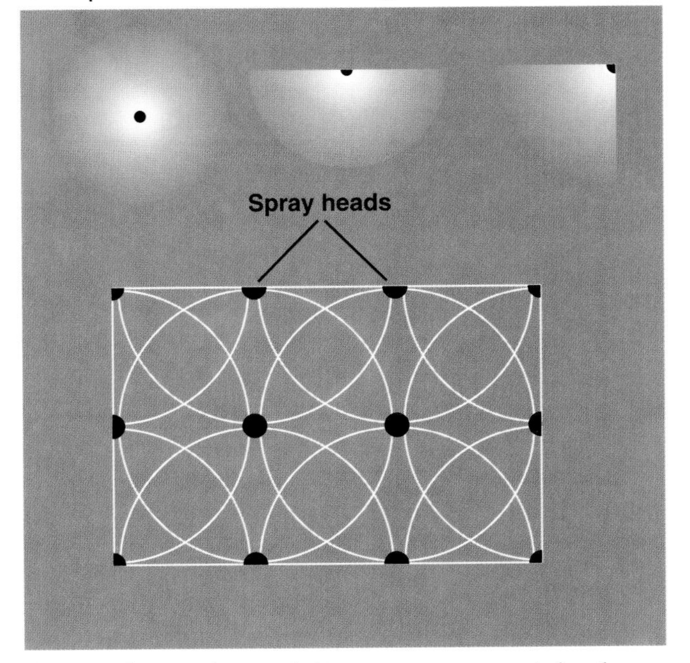

System layout is crucial to ensure proper irrigation of all parts of your yard. Spray heads are available in full-circle, semicircle, or quarter-circle patterns to cover the entire space. In most cases, you will divide your landscape into several individual zones, each controlled by its own valve. With a timer (photo below), you can program precise start and stop times for each irrigation zone.

Photo courtesy of Hunter Industries

Valve manifold is a group of valves used to control the various sprinkler zones. Some models are installed below ground in a valve box, while others extend above the ground. When a timer is used, each control valve in the manifold is wired separately into the timer.

Photo courtesy of Hunter Industries

Sprinkler timers can be programmed to provide automatic control of all zones in an underground sprinkler system. Deluxe models control up to 16 different zones and have rain sensors that shut off the system when irrigation is not needed.

Photo courtesy of Hunter Industries

Sprinkler heads come in many styles to provide a variety of spray patterns. Flexible underground tubing, sometimes called *funny pipe*, links the spray head to saddle fittings on the underground water supply pipes.

Replacing Old Plumbing

Leave old plumbing pipes in place, if possible. To save time, professional plumbing contractors remove old plumbing pipes only when they interfere with the routing of the new plumbing lines.

Replacing Old Plumbing

Plumbing pipes, like all building materials, eventually wear out and have to be replaced. If you find yourself repairing leaky, corroded pipes every few months, it may be time to consider replacing the old system entirely—and soon. A corroded water pipe that bursts while you are away can cost you many thousands of dollars in damage to wall surfaces, framing members, and furnishings.

Identifying the materials used in your plumbing system can also tell you if replacement is advised. If you have galvanized steel pipes, for example, it is a good bet that they will need to be replaced in the near future. Most galvanized steel pipes were installed before 1960, and since steel pipes have a maximum life expectancy of 30 to 35 years, such a system is probably living on borrowed time. On the other hand, if your system includes copper supply pipes and plastic drain pipes, you can relax; these materials were likely installed within the last 30 years, and they are considerably more durable than steel, provided they were installed correctly.

Unless you live in a rambler with an exposed basement ceiling, replacing old plumbing nearly always involves some demolition and carpentry work. Even in the best scenario, you probably will find it necessary to open walls and floors in order to run new pipes. For this reason, replacing old plumbing is often done at the same time as a kitchen or bathroom remodeling project, when wall and floor surfaces have to be removed and replaced.

A plumbing renovation project is subject to the same Code regulations as a new installation. Always work in conjunction with your local inspector (pages 34 to 39) when replacing old plumbing.

This section shows:

- Evaluating Your Plumbing (pages 88 to 89)
- Step-by-step Overview (pages 90 to 91)
- Planning Pipe Routes (pages 92 to 95)
- Replacing a Main Waste-Vent Stack (pages 96 to 101)
- Replacing Branch Drains & Vent Pipes (pages 102 to 105)
- Replacing a Toilet Drain (pages 106 to 107)
- Replacing Supply Pipes (pages 108 to 109)

Replacement Options

Partial replacement involves replacing only those sections of your plumbing system that are currently causing problems. This is a quick, less expensive option than a complete renovation, but it is only a temporary solution. Old plumbing will continue to fail until you replace the entire system.

Complete replacement of all plumbing lines is an ambitious job, but doing this work yourself can save you thousands of dollars. To minimize the inconvenience, you can do this work in phases, replacing one branch of the plumbing system at a time.

Evaluating Your Plumbing

By the time you spot the telltale evidence of a leaky drain pipe or water supply pipe, the damage to the walls and ceilings of your home can be considerable. The tips on the following pages show early warning signals that indicate your plumbing system is beginning to fail.

Proper evaluation of your plumbing helps you identify old, suspect materials and anticipate problems. It also can save you money and aggravation. Replacing an old plumbing system at your convenience before it reaches the disaster stage is preferable to hiring a plumbing contractor to bail you out of an emergency situation.

Remember that the network of pipes running through the walls of your home is only one part of the larger system. You should also evaluate the main water supply and sewer pipes that connect your home to the city utility system and make sure they are adequate before you replace your plumbing.

Fixture Units	Minimum Gallons per Minute (GPM)
10	8
15	11
20	14
25	17
30	20

Minimum recommended water capacity is based on total demand on the system, as measured by fixture units, a standard of measurement assigned by the Plumbing Code. First, add up the total units of all the fixtures in your plumbing system (page 36). Then, perform the water supply capacity test described below. Finally, compare your water capacity with the recommended minimums listed above. If the capacity falls below that recommended in the table above, then the main water supply pipe running from the city water main to your home is inadequate and should be replaced with a larger pipe by a licensed contractor.

How to Determine Your Water Supply Capacity

1 Shut off the water at the valve on your water meter, then disconnect the pipe on the house side of the meter. Construct a test spout using a 2" PVC elbow and two 6" lengths of 2" PVC pipe, then place the spout on the exposed outlet on the water meter. Place a large watertight tub under the spout to collect water.

2 Open the main supply valve and let the water run into the container for 30 seconds. Shut off the water, then measure the amount of water in the container and multiply this figure by 2. This number represents the gallons-per-minute rate of your main water supply. Compare this measurement with the recommended capacity in the table above.

Symptoms of Bad Plumbing

Rust stains on the surfaces of toilet bowls and sinks may indicate severe corrosion inside iron supply pipes. This symptom generally means your water supply system is likely to fail in the near future. NOTE: Rust stains can also be caused by a water heater problem or by a water supply with a high mineral content. Check for these problems before assuming your pipes are bad.

Low water pressure at fixtures suggests that the supply pipes either are badly clogged with rust and mineral deposits, or are undersized. To measure water pressure, plug the fixture drain and open the faucets for 30 seconds. Measure the amount of water and multiply by 2; this figure is the gallons-per-minute rating. Vanity faucets should supply 1¾ gpm; bathtub faucets, 6 gpm; kitchen sink faucets, 4½ gpm.

Slow drains throughout the house may indicate that DWV pipes are badly clogged with rust and mineral deposits. When a fixture faucet is opened fully with the drains unstopped, water should not collect in tubs and basins. NOTE: Slow drains may also be the result of inadequate venting. Check for this problem before assuming the drain pipes are bad.

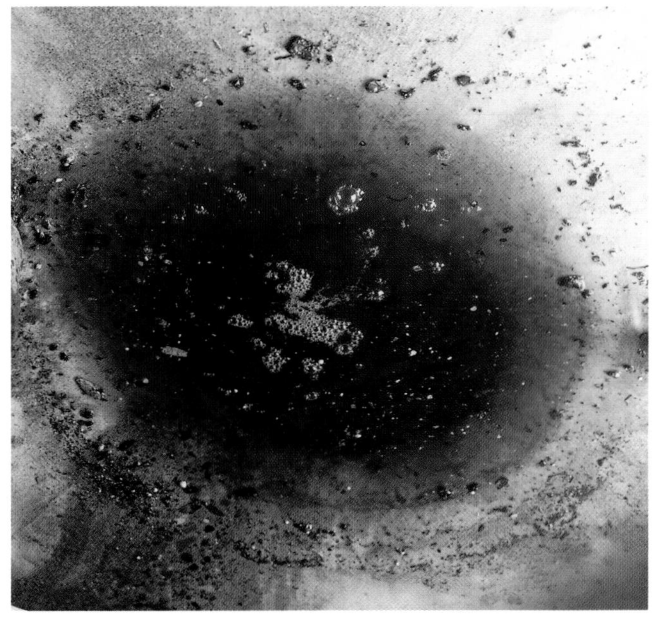

Backed-up floor drains indicate that the main sewer service to the street is clogged. If you have this problem regularly, have the main sewer lines evaluated by a plumbing contractor before you replace your house plumbing. The contractor will be able to determine if your sewer problem is a temporary clog or a more serious problem that requires major work.

Replacing Old Plumbing: A Step-by-step Overview

The overview sequence shown here represents the basic steps you will need to follow when replacing DWV and water supply pipes. On the following pages, you will see these steps demonstrated in complete detail, as we replace all the water supply pipes and drain pipes for a bathroom, including the main waste-vent stack running from basement to roofline.

Remember that no two plumbing jobs are ever alike, and your own project will probably differ from the demonstration projects shown in this section. Always work in conjunction with your local plumbing inspector, and organize your work around a detailed plumbing plan that shows the particulars of your project. Review pages 8 to 41 of this book before starting work.

1 Plan the routes for the new plumbing pipes. Creating efficient pathways for new pipes is crucial to a smooth installation. In some cases this requires removing wall or floor surfaces. Or, you can frame a false wall, called a *chase* (page 92), to create space for running the new pipes.

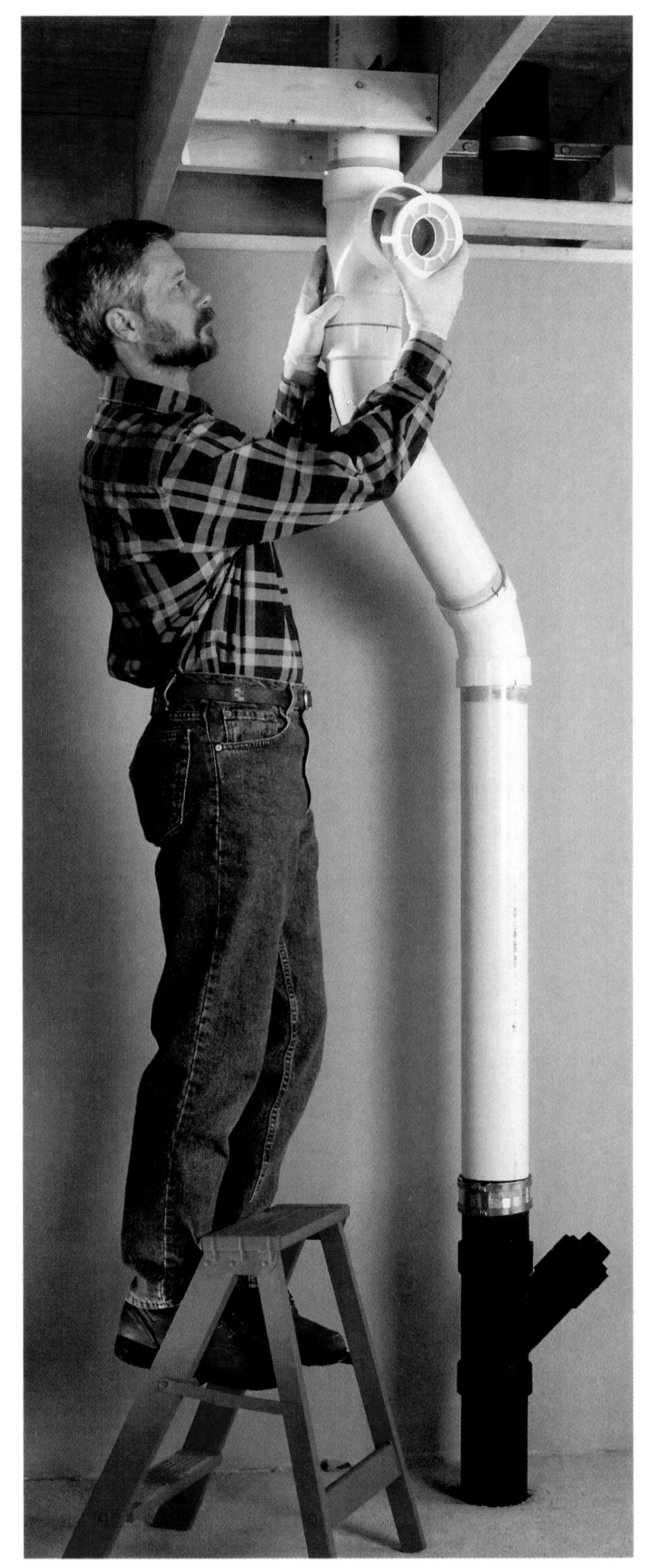

2 Remove sections of the old waste-vent stack, as needed, then install a new main waste-vent stack running from the main drain in the basement to the roof. Include the fittings necessary to connect branch drains and vent lines to the stack.

3 Install new branch drains from the waste-vent stack to the stub-outs for the individual fixtures. If the fixture locations have not changed, you may need to remove sections of the old drain pipes in order to run the new pipes.

4 Remove the old toilet bend and replace it with a new bend running to the new waste-vent stack. This task usually requires that you remove areas of flooring. Framing work may also be required to create a path for the toilet drain.

5 Replace the vent pipes running from the fixtures up to the attic, then connect them to the new waste-vent stack.

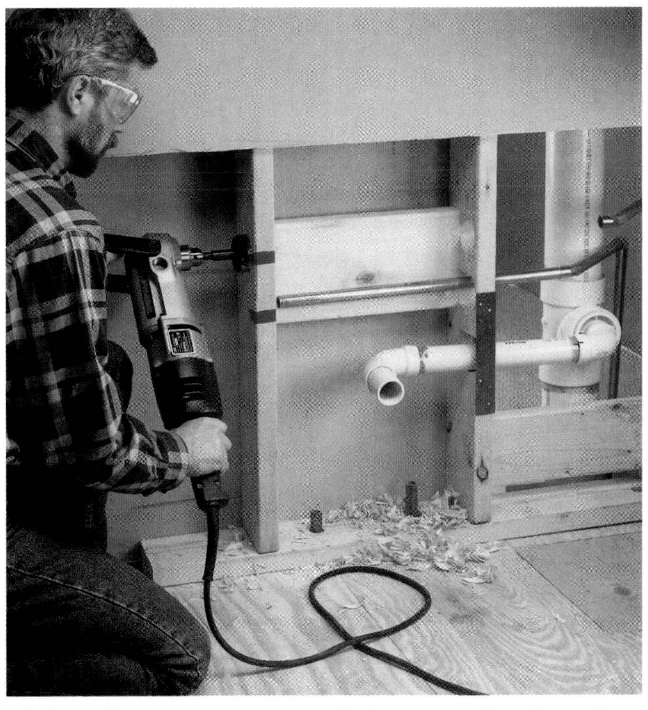

6 Install new copper supply lines running from the water meter to all fixture locations. Test the DWV and water supply pipes and have your work inspected before closing up walls and installing the fixtures.

Build a framed chase. A chase is a false wall created to provide space for new plumbing pipes. It is especially effective for installing a new main waste-vent stack. On a two-story house, chases can be stacked one over the other on each floor in order to run plumbing from the basement to the attic. Once plumbing is completed and inspected, the chase is covered with wallboard and finished to match the room.

Planning Pipe Routes

The first, and perhaps most important, step when replacing old plumbing is to decide how and where to run the new pipes. Since the stud cavities and joist spaces are often covered with finished wall surfaces, finding routes for running new pipes can be challenging.

When planning pipe routes, choose straight, easy pathways whenever possible. Rather than running water supply pipes around wall corners and through studs, for example, it may be easiest to run them straight up wall cavities from the basement. Instead of running a bathtub drain across floor joists, run it straight down into the basement, where the branch drain can be easily extended underneath the joists to the main waste-vent stack.

In some situations, it is most practical to route the new pipes in wall and floor cavities that already hold plumbing pipes, since these spaces often are framed to provide long, unobstructed runs. A detailed map of your plumbing system can be very helpful when planning routes for new plumbing pipes (pages 14 to 19).

To maximize their profits, plumbing contractors generally try to avoid opening walls or changing wall framing when installing new plumbing. But the do-it-yourselfer does not have these limitations. Faced with the difficulty of running pipes through enclosed spaces, you may find it easiest to remove wall surfaces or to create a newly framed space for running new pipes.

On these pages, you will see some common methods used to create pathways for replacing old pipes with new plumbing.

Tips for Planning Pipe Routes

Use existing access panels to disconnect fixtures and remove old pipes. Plan the location of new fixtures and pipe runs to make use of existing access panels, minimizing the amount of demolition and repair work you will need to do.

Convert a laundry chute into a channel for running new plumbing pipes. The door of the chute can be used to provide access to control valves, or it can be removed and covered with wall materials, then finished to match the surrounding wall.

Run pipes inside a closet. If they are unobtrusive, pipes can be left exposed at the back of the closet. Or, you can frame a chase to hide the pipes after the installation is complete.

Remove false ceiling panels to route new plumbing pipes in joist cavities. Or, you can route pipes across a standard plaster or wallboard ceiling, then construct a false ceiling to cover the installation, provided there is adequate height. Most Building Codes require a minimum of 7 ft. from floor to finished ceiling.

(continued next page)

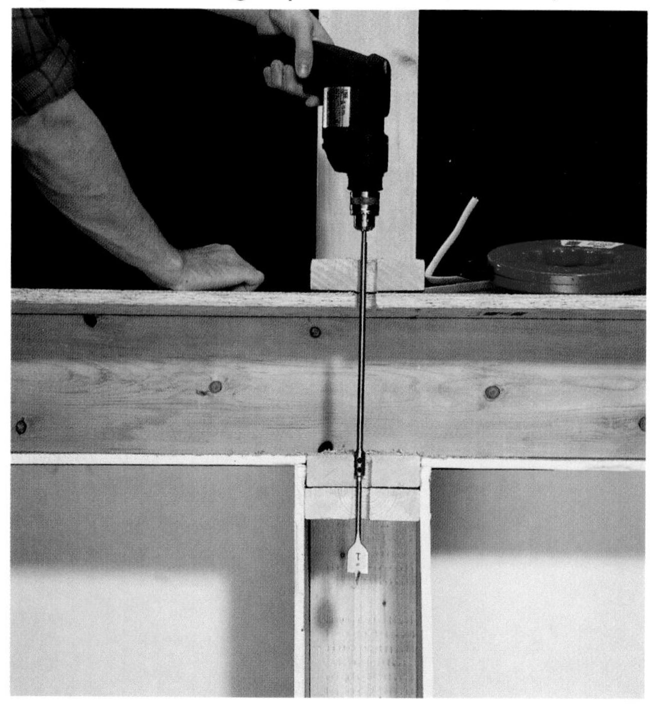

Use a drill extension and spade bit or hole saw to drill through wall plates from unfinished attic or basement spaces above or below the wall.

Look for "wet walls." Walls that hold old plumbing pipes can be good choices for running long vertical runs of new pipe. These spaces usually are open, without obstacles such as fireblocks and insulation.

Probe wall and floor cavities with a long piece of plastic pipe to ensure that a clear pathway exists for running new pipe (left). Once you have established a route using the narrow pipe, you can use the pipe as a guide when running larger drain pipes up into the wall (right).

Remove flooring when necessary. Because replacing toilet and bathtub drains usually requires that you remove sections of floor, a full plumbing replacement job is often done in conjunction with a complete bathroom remodeling project.

Remove wall surfaces when access from above or below the wall is not possible. This demolition work can range from cutting narrow channels in plaster or wallboard to removing the entire wall surface. Remove wall surfaces back to the centers of adjoining studs; the exposed studs provide a nailing surface for attaching new wall materials once the plumbing project is completed.

Create a detailed map showing the planned route for your new plumbing pipes. Such a map can help you get your plans approved by the inspector, and it makes work much simpler. If you have already mapped your existing plumbing system (pages 14 to 19), those drawings can be used to plan new pipe routes.

A new main waste-vent stack is best installed near the location of the old stack. In this way, the new stack can be connected to the basement floor cleanout fitting used by the old cast-iron stack.

Replacing a Main Waste-Vent Stack

Although a main waste-vent stack rarely rusts through entirely, it can be nearly impossible to join new branch drains and vents to an old cast-iron stack. For this reason, plumbing contractors sometimes recommend replacing the iron stack with plastic pipe during a plumbing renovation project.

Be aware that replacing a main waste-vent stack is not an easy job (pages 32 to 33). You will be cutting away heavy sections of cast iron, so working with a helper is essential. Before beginning work, make sure you have a complete plan for your plumbing system and have designed a stack that includes all the fittings you will need to connect branch drains and vent pipes. While work is in progress, none of your plumbing fixtures will be usable. To speed up the project and minimize inconvenience, do as much of the demolition and preliminary construction work as you can before starting work on the stack.

Because main waste-vent stacks may be as large as 4" in diameter, running a new stack through existing walls can be troublesome. To solve this problem, our project employs a common solution: framing a chase in the corner of a room to provide the necessary space for running the new stack from the basement to the attic. When the installation is complete, the chase will be finished with wallboard to match the room.

How to Replace a Main Waste-Vent Stack

1 Secure the cast-iron waste-vent stack near the ceiling of your basement, using a riser clamp installed between the floor joists. Use wood blocks attached to the joists with 3" wallboard screws to support the clamp. Also clamp the stack in the attic, at the point where the stack passes down into the wall cavity. WARNING: A cast-iron stack running from basement to attic can weigh several hundred pounds. Never cut into a cast-iron stack before securing it with riser clamps above the cut.

2 Use a cast iron snap cutter to sever the stack near the floor of the basement, about 8" above the cleanout, and near the ceiling, flush with the bottom of the joists. Have a helper hold the stack while you are cutting out the section. NOTE: After cutting into the main waste-vent stack, plug the open end of the pipe with a cloth to prevent sewer gases from rising into your home.

3 Nail blocking against the bottom of the joists across the severed stack. Then, cut a 6"-diameter hole in the basement ceiling where the new waste-vent stack will run, using a reciprocating saw. Suspend a plumb bob at the centerpoint of the opening as a guide for aligning the new stack.

(continued next page)

4 Attach a 5-ft. segment of PVC plastic pipe the same diameter as the old waste-vent stack to the exposed end of the cast-iron cleanout fitting, using a banded coupling with neoprene sleeve (page 33).

5 Dry-fit 45° elbows and straight lengths of plastic pipe to offset the new stack, lining it up with the plumb bob centered on the ceiling opening.

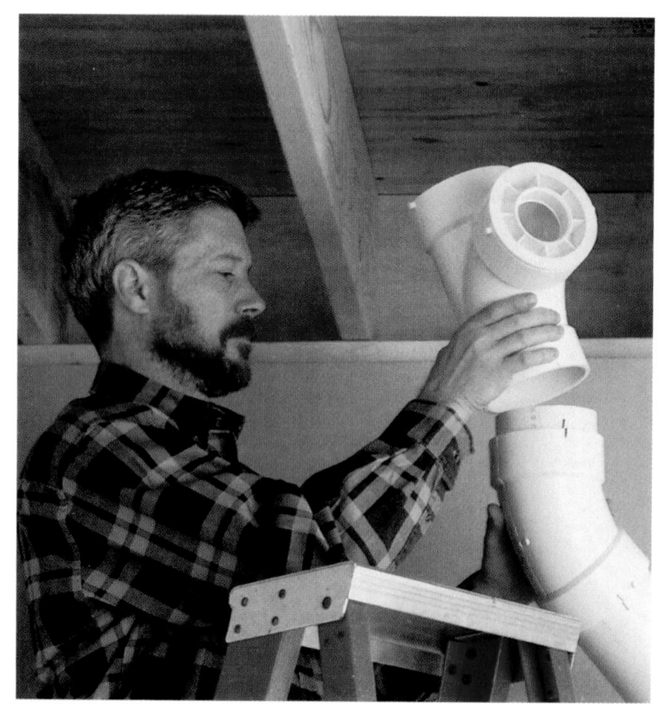

6 Dry-fit a waste T-fitting on the stack, with the inlets necessary for any branch drains that will be connected in the basement. Make sure the fitting is positioned at a height that will allow the branch drains to have the correct ¼" per foot downward slope toward the stack.

7 Determine the length for the next piece of waste-vent pipe by measuring from the basement T-fitting to the next planned fitting in the vertical run. In our project, we will be installing a T-fitting between floor joists, where the toilet drain will be connected.

8 Cut a PVC plastic pipe to length, raise it into the opening, and dry-fit it to the T-fitting. NOTE: For very long pipe runs, you may need to construct this vertical run by solvent-gluing two or more segments of pipe together with couplings.

9 Check the length of the stack, then solvent-glue all fittings together. Support the new stack with a riser clamp resting on blocks attached between basement ceiling joists.

10 Attach the next waste T-fitting to the stack. In our demonstration project, the waste T lies between floor joists and will be used to connect the toilet drain. Make sure the waste T is positioned at a height which will allow for the correct ⅛" per foot downward slope for the toilet drain.

11 Add additional lengths of pipe, with waste T-fittings installed where other fixtures will drain into the stack. In our example, a waste T with a 1½" bushing insert is installed where the vanity sink drain will be attached to the stack. Make sure the T-fittings are positioned to allow for the correct downward pitch of the branch drains.

(continued next page)

12 Cut a hole in the ceiling where the waste-vent stack will extend into the attic, then measure, cut, and solvent-glue the next length of pipe in place. The pipe should extend at least 1 ft. up into the attic.

13 Remove the roof flashing from around the old waste-vent stack. You may need to remove shingles in order to accomplish this. NOTE: Always use caution when working on a roof. If you are unsure of your ability to do this work, hire a roof repair specialist to remove the old flashing and install new flashing around the new vent pipe.

14 In the attic, remove old vent pipes, where necessary, then sever the cast-iron soil stack with a cast iron cutter and lower the stack down from the roof opening with the aid of a helper. Support the old stack with a riser clamp installed between joists.

15 Solvent-glue a vent T with a 1½" bushing in the side inlet to the top of the new waste-vent stack. The side inlet should point toward the nearest auxiliary vent pipe extending up from below.

16 Finish the waste-vent stack installation by using 45° elbows and straight lengths of pipe to extend the stack through the same roof opening used by the old vent stack. The new stack should extend at least 1 ft. through the roof, but no more than 2 ft.

How to Flash a Waste-Vent Stack

1 Loosen the shingles directly above the new vent stack, and remove any nails, using a flat pry bar. When installed, the metal vent flashing will lie flat on the shingles surrounding the vent pipe. Apply roofing cement to the underside of the flashing.

2 Slide the flashing over the vent pipe, and carefully tuck the base of the flashing up under the shingle. Press the flange firmly against the roof deck to spread the roofing cement, then anchor it with rubber gasket flashing nails. Reattach loose shingles as necessary.

Remove old pipes only where they obstruct the planned route for the new pipes. You will probably need to remove drain and water supply pipes at each fixture location, but the remaining pipes usually can be left in place. A reciprocating saw with metal-cutting blade works well for this job.

Replacing Branch Drains & Vent Pipes

In our demonstration project, we are replacing branch drains for a bathtub and vanity sink. The tub drain will run down into the basement before connecting to the main waste-vent stack, while the vanity drain will run horizontally to connect directly to the stack.

A vent pipe for the bathtub runs up into the attic, where it will join the main waste-vent stack. The vanity sink, however, requires no secondary vent pipe, since its location falls within the critical distance (page 39) of the new waste-vent stack.

How to Replace Branch Drains

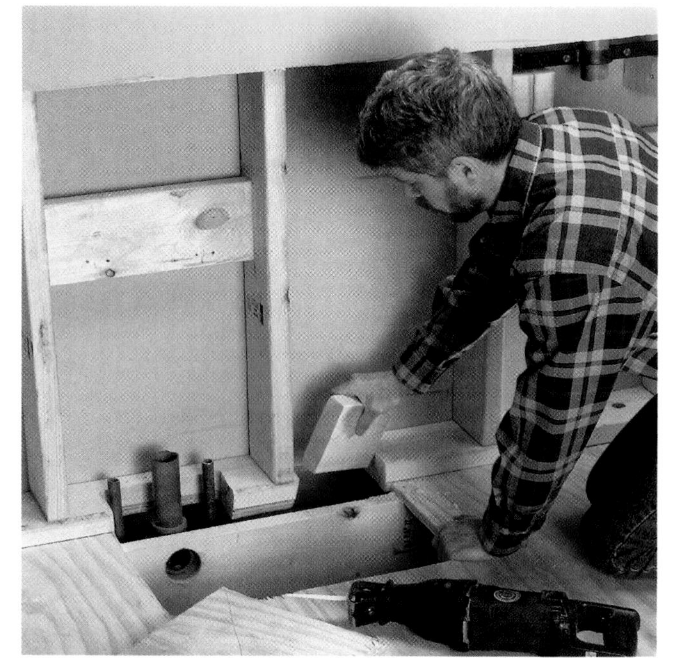

1 Establish a route for vertical drain pipes running through wall cavities down into the basement. For our project, we are cutting away a section of the wall sole plate in order to run a 1½" bathtub drain pipe from the basement up to the bathroom.

2 From the basement, cut a hole in the bottom of the wall, below the opening cut (step 1). Measure, cut, and insert a length of vertical drain pipe up into the wall to the bathroom. A length of flexible CPVC pipe can be useful for guiding the drain pipe up into the wall. For very long pipe runs, you may need to join two or more lengths of pipe with couplings as you insert the run.

3 Secure the vertical drain pipe with a riser clamp supported on 2 × 4 blocks nailed between joists. Take care not to overtighten the clamps.

4 Install a horizontal pipe from the waste T-fitting on the waste-vent stack to the vertical drain pipe. Maintain a downward slope toward the stack of ¼" per foot, and use a Y-fitting with 45° elbow to form a cleanout where the horizontal and vertical drain pipes meet.

5 Solvent-glue a waste T-fitting to the top of the vertical drain pipe. For a bathtub drain, as shown here, the T-fitting must be well below floor level to allow for the bathtub drain trap (page 113). You may need to notch or cut a hole in floor joists to connect the drain trap to the waste T (page 46).

6 From the attic, cut a hole into the top of the bathroom wet wall, directly above the bathtub drain pipe. Run a 1½" vent pipe down to the bathtub location, and solvent-glue it to the waste T. Make sure the pipe extends at least 1 ft. into the attic.

(continued next page)

7 Remove wall surfaces as necessary to provide access for running horizontal drain pipes from fixtures to the new waste-vent stack. In our project, we are running 1½" drain pipe from a vanity sink to the stack. Mark the drain route on the exposed studs, maintaining a ¼" per foot downward slope toward the stack. Use a reciprocating saw or jig saw to notch out the studs (page 46).

8 Secure the old drain and vent pipes with riser clamps supported by blocking attached between the studs.

9 Remove the old drain and water supply pipes, where necessary, to provide space for running the new drain pipes.

10 Using a sweep elbow and straight length of pipe, assemble a drain pipe to run from the drain stub-out location to the waste T-fitting on the new waste-vent stack. Use a 90° elbow and a short length of pipe to create a stub-out extending at least 2" out from the wall. Secure the stub-out to a ¾" backer board attached between studs.

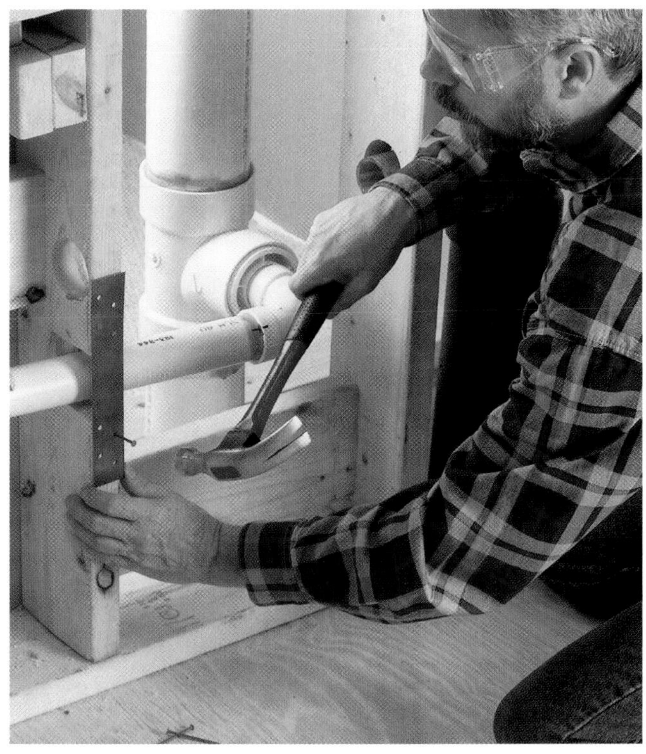

11 Protect the drain pipes by attaching metal protector plates over the notches in the studs. Protector plates prevent drain pipes from being punctured when wall surfaces are replaced.

12 In the attic, use a vent elbow and straight length of pipe to connect the vertical vent pipe from the tub to the new waste-vent stack.

Replacing a toilet drain usually requires that you remove flooring and wall surface to gain access to the pipes.

Replacing a Toilet Drain

Replacing a toilet drain is sometimes a troublesome task, mostly because the cramped space makes it difficult to route the large, 3" or 4" pipe. You likely will need to remove flooring around the toilet and wall surface behind the toilet.

Replacing a toilet drain may require framing work, as well, if you find it necessary to cut into joists in order to route the new pipes. When possible, plan your project to avoid changes to the framing members.

How to Replace a Toilet Drain

1 Remove the toilet, then unscrew the toilet flange from the floor and remove it from the drain pipe. NOTE: If the existing toilet flange is cast iron or bronze, it may be joined to the toilet bend with poured lead or solder; in this case, it is easiest to break up the flange with a masonry hammer (make sure to wear eye protection) and remove it in pieces.

2 Cut away the flooring around the toilet drain along the center of the floor joists, using a circular saw with the blade set to a depth ⅛" more than the thickness of the subfloor. The exposed joist will serve as a nailing surface when the subfloor is replaced.

3 Cut away the old toilet bend as close as possible to the old waste-vent stack, using a reciprocating saw with metal-cutting blade, or a cast iron cutter.

4 If a joist obstructs the route to the new waste-vent stack, cut away a section of the floor joist. Install double headers and metal joist hangers to support the ends of the cut joist.

5 Create a new toilet drain running to the new waste-vent stack, using a toilet bend and a straight length of pipe. Position the drain so there will be at least 15" of space between the center of the bowl and side wall surfaces when the toilet is installed. Make sure the drain slopes at least ⅛" per foot toward the stack, then support the pipe with plastic pipe strapping attached to framing members. Insert a 6" length of pipe in the top inlet of the closet bend; once the new drain pipes have been tested, this pipe will be cut off with a handsaw and fitted with a toilet flange.

6 Cut a piece of exterior-grade plywood to fit the cutout floor area, and use a jig saw to cut an opening for the toilet drain stub-out. Position the plywood, and attach it to joists and blocking with 2" screws.

Support copper supply pipes every 6 ft. along vertical runs and 10 ft. along horizontal runs. Always use copper or plastic support materials with copper; never use steel straps, which can interact with copper and cause corrosion.

Replacing Supply Pipes

When replacing old galvanized water supply pipes, we recommend that you use type-M rigid copper. Use ¾" pipe for the main distribution pipes and ½" pipes for the branch lines running to individual fixtures.

For convenience, run hot and cold water pipes parallel to one another, between 3" and 6" apart. Use the straightest, most direct routes possible when planning the layout, because too many bends in the pipe runs can cause significant friction and reduce water pressure.

It is a good idea to removed old supply pipes that are exposed, but pipes hidden in walls can be left in place unless they interfere with the installation of the new supply pipes.

How to Replace Water Supply Pipes

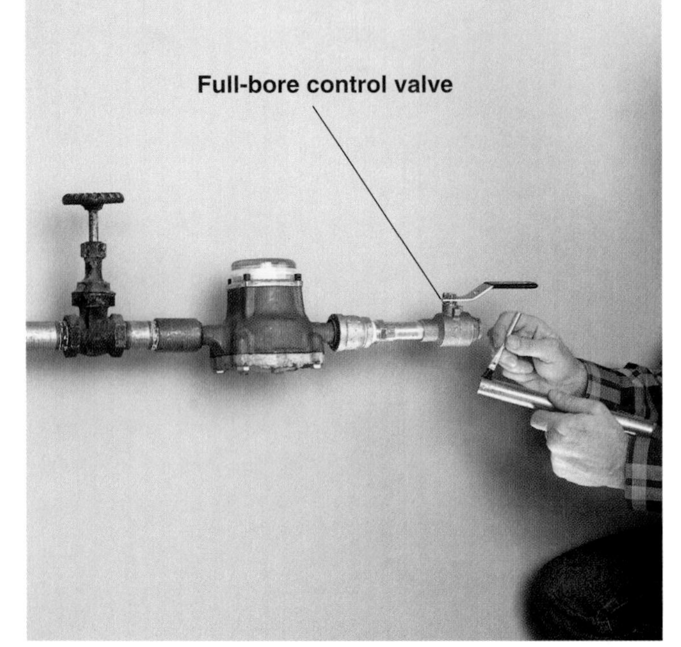

1 Shut off the water on the street side of the water meter, then disconnect and remove the old water pipes from the house side. Solder a ¾" male threaded adapter and full-bore control valve to a short length of ¾" copper pipe, then attach this assembly to the house side of the water meter. Extend the ¾" cold-water distribution pipe toward the nearest fixture, which is usually the water heater.

2 At the water heater, install a ¾" T-fitting in the cold-water distribution pipe. Use two lengths of ¾" copper pipe and a full-bore control valve to run a branch pipe to the water heater. From the outlet opening on the water heater, extend a ¾" hot-water distribution pipe, also with a full-bore control valve (page 117). Continue the hot and cold supply lines on parallel routes toward the next group of fixtures in your house.

3 Establish routes for branch supply lines by drilling holes into stud cavities. Install T-fittings, then begin the branch lines by installing brass control valves. Branch lines should be made with ¾" pipe if they are supplying more than one fixture; ½" if they are supplying only one fixture.

4 Extend the branch lines to the fixtures. In our project, we are running ¾" vertical branch lines up through the framed chase to the bathroom. Route pipes around obstacles, such as a main waste-vent stack, by using 45° and 90° elbows and short lengths of pipe.

5 Where branch lines run through studs or floor joists, drill holes or cut notches in the framing members (page 46), then insert the pipes. For long runs of pipe, you may need to join two or more shorter lengths of pipe, using couplings as you create the runs.

6 Install ¾" to ½" reducing T-fittings and elbows to extend the branch lines to individual fixtures. In our bathroom, we are installing hot and cold stub-outs for the bathtub and sink, and a cold-water stub-out for the toilet. Cap each stub-out until your work has been inspected and the wall surfaces have been completed.

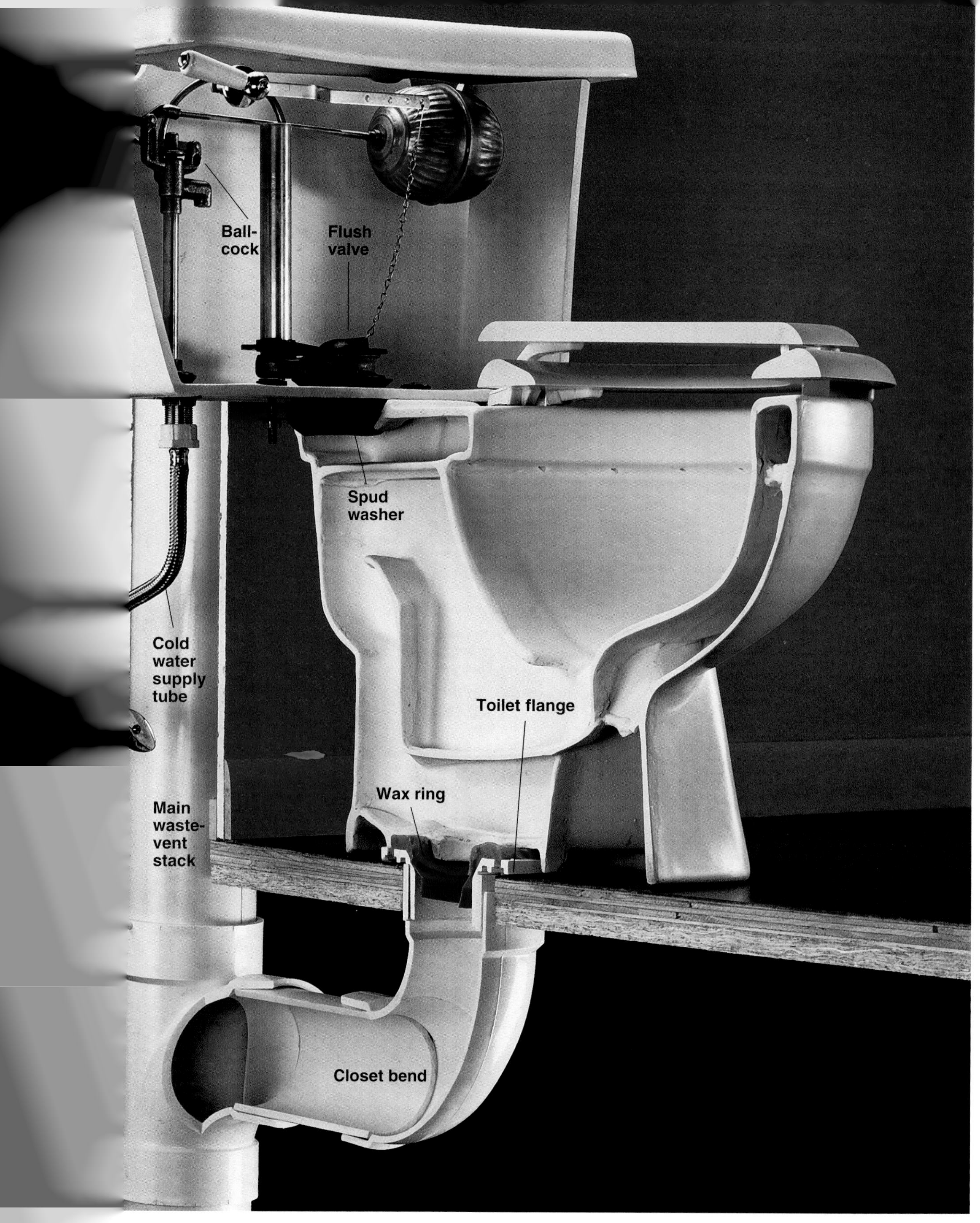

e connected to a cold water supply stub-
ans of a supply tube running from the shut-
o the ballcock fitting on the bottom of the
. This supply tube can be made of steel
omed copper, vinyl, or plastic mesh. The
is mounted to the toilet base with tank
the opening between tank and base is sealed with a rubber spud washer. The drain is connected by a toilet flange attached to the top of the closet bend. A wax ring with a rubber sleeve fits between the toilet drain opening (called the *horn*) and the toilet flange. Floor bolts attached to the flange fit through openings in the toilet base and are used to tighten the toilet down against the floor.

Making Final Connections

The photos on the following pages show the basics for making the final connections between your new DWV and water supply lines and the fixtures in your new plumbing project. Remember, however, that fixture connections vary widely. Always follow manufacturers' directions. For additional information, you can consult other books on plumbing and remodeling (right). In this final section, you will find information on connecting:

- Toilet (opposite page)
- Vanity sink, faucet (below)
- Shower (page 112)
- Bathtub drain, faucet (page 113)
- Kitchen sink (page 114)
- Dishwasher (page 114)
- Sillcock (116)
- Kitchen faucet (page 116)
- Refrigerator icemaker (page 116)
- Water heater (page 117)

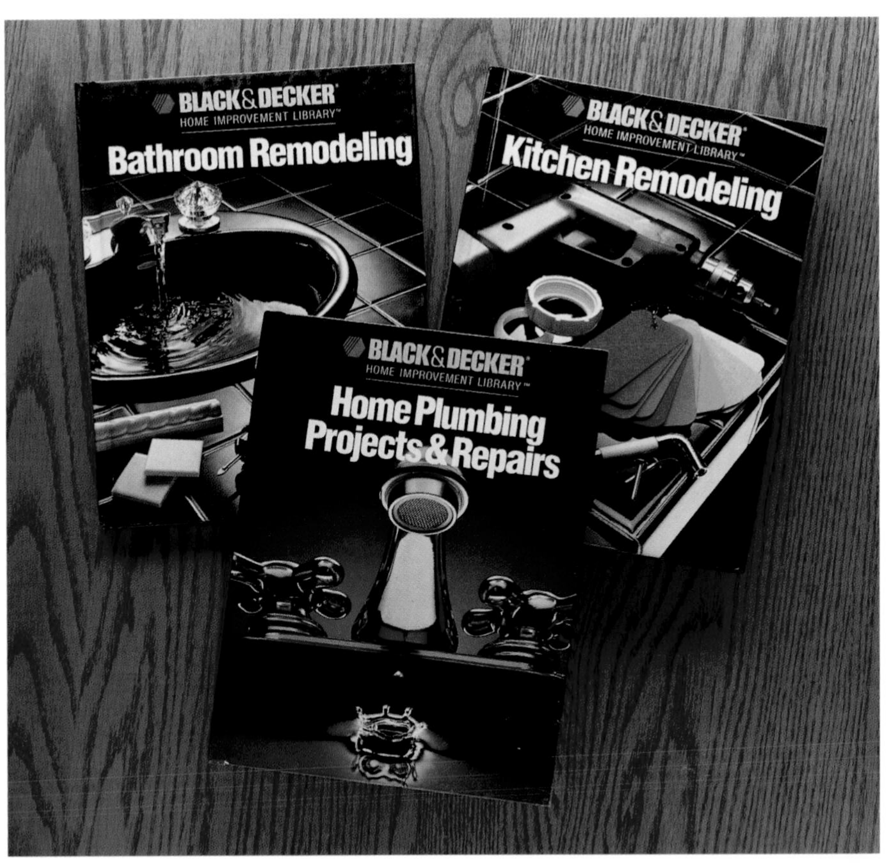

Other sources of information include these books from the Black & Decker® Home Improvement Library™: *Home Plumbing Projects & Repairs*, *Bathroom Remodeling*, and *Kitchen Remodeling*. Books like these can guide you when choosing fixtures, as well as provide valuable installation information.

Bathroom sink drain is connected with a P-trap that links the sink tailpiece to the branch drain stub-out. Drains for sinks use traps secured with slip fittings, so that the trap can be removed to provide access when servicing the drain.

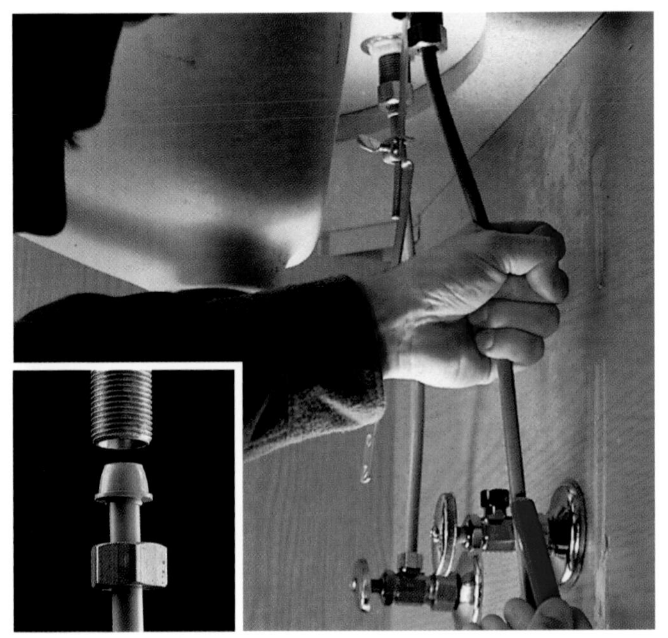

Bathroom sink faucet connects to the water supply stub-outs with flexible supply tubes that run from the shutoff valves to the faucet tailpieces. Coupling nuts secure the tubes to the shutoff valves and to the tailpieces (inset).

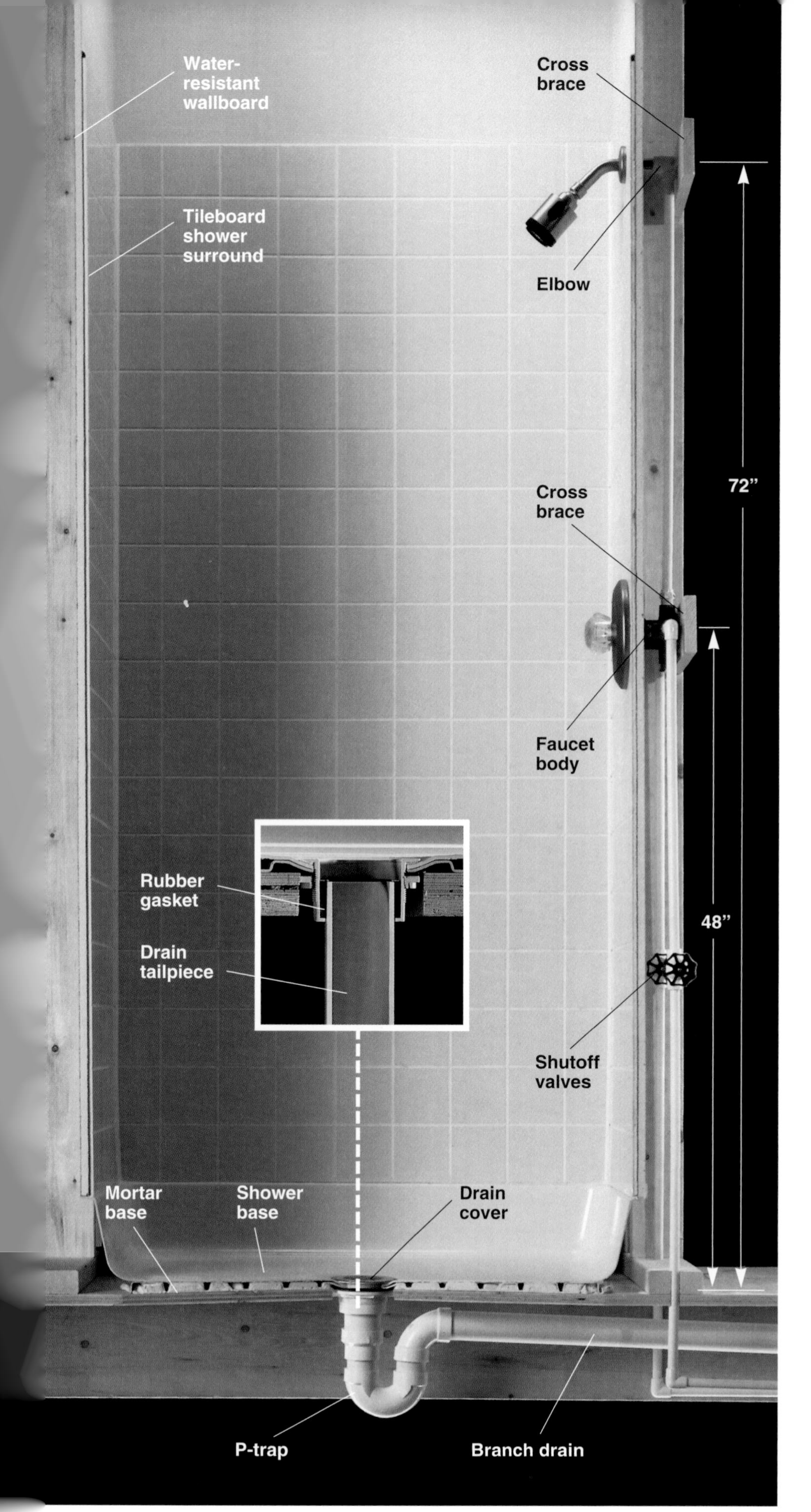

Showers are usually contained in a framed stall constructed with 2 ¥ 4s. The shower stall can be lined with ceramic tile, or it can be constructed using a prefabicated fiberglass shower surround and shower base, as shown here. The "wet wall" containing the shower supply pipes should have an access panel so you can reach shutoff valves when servicing the shower.

When the shower is constructed with a purchased shower surround and base, a layer of mortar is first applied to the subfloor to create a bed for the shower base. Water-resistant wallboard or cementboard is then attached to the stall framing members to provide a surface for gluing the shower panels in place.

Water supply connections: Water supply risers from the hot and cold branch pipes extend up through the floor and are connected to shutoff valves. From the shutoff valves, risers run up to the faucet body. From the faucet mixing chamber, a single shower pipe leads up to the shower arm and shower head extending through the wall.

Drain connections: The shower base is positioned so the drain cover assembly slides over the drain tailpiece. A P-trap is attached to the branch drain and tailpiece. If the shower drain is accessible, it should use a detachable trap joined with slip nuts. If the drain in not accessible, it should be formed with permanent solvent-glued components.

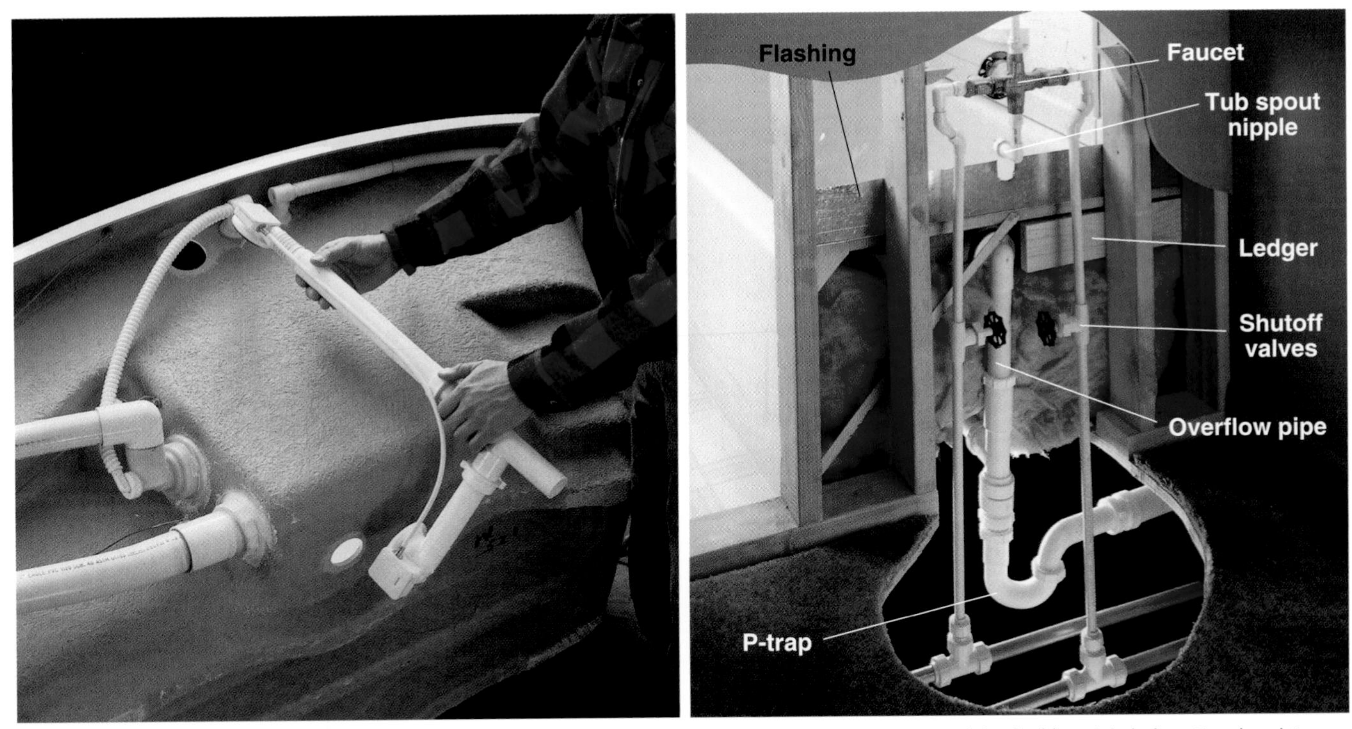

Bathtubs are connected to the drain system with a drain-waste-overflow assembly (left), which is attached to the tub before it is set into position. Most whirlpool tubs come with this assembly, but with standard tubs you usually must purchase a drain-waste-overflow kit. The bathtub faucet set is connected by running risers from the branch hot and cold water lines to the faucet inlets. Shutoff valves are installed to control each riser (right).

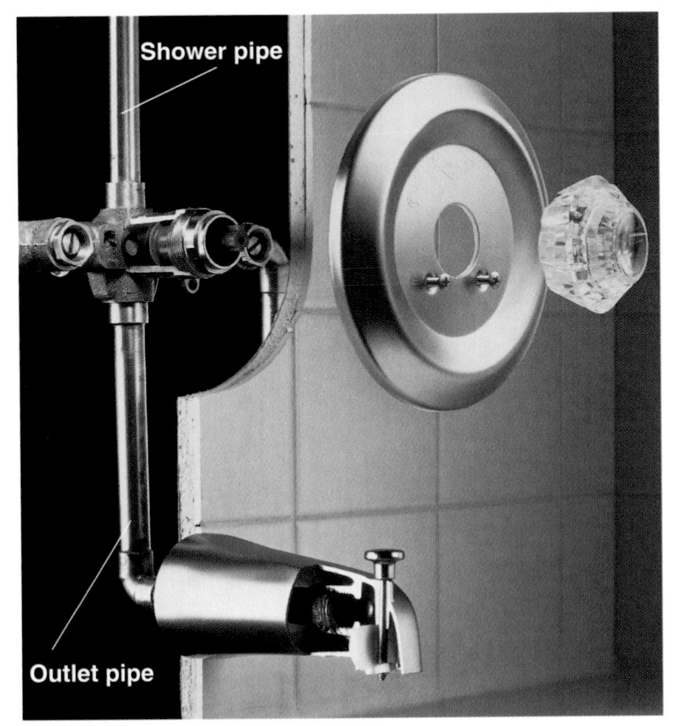

Bathtub faucets are typically embedded inside wall cavities and are usually anchored to 2 × 4 blocking installed between wall studs. An outlet pipe attached to the faucet body extends through the wall and has a threaded fitting to accept a spout. Another faucet outlet can be used to extend a pipe up the wall to the shower head.

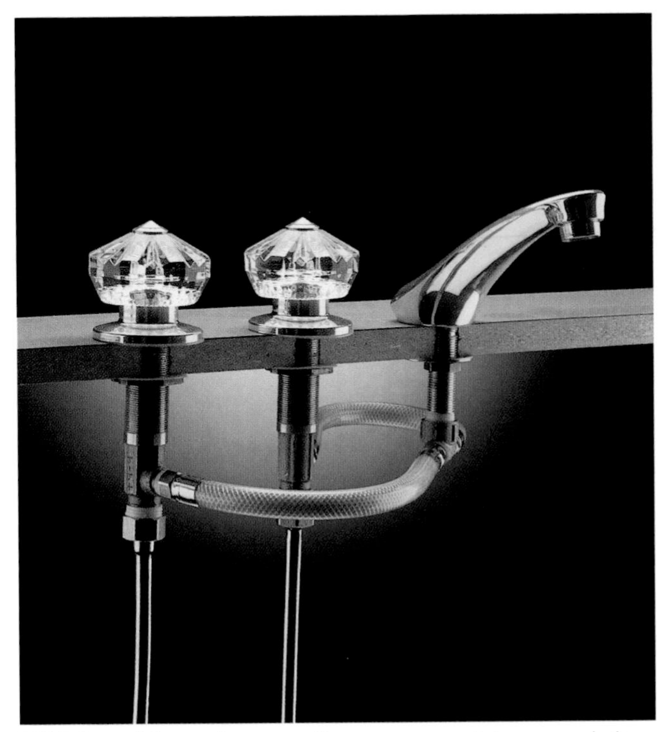

Whirlpool faucets sometimes use a widespread design, with handles that are mounted separately from the spout on a platform surrounding the tub. Flex tubes available in many lengths make it possible to mount the spout a considerable distance from the faucet valves.

Kitchen sink with two basins is connected to the drain system by means of a P-trap and continuous waste T-fitting. One end of the continuous waste T connects to the food disposer, the other end to a drain tailpiece running to the other sink basin. Supply tubes controlled by shutoff valves bring water from the hot and cold supply risers to the faucet. The hot water riser may have a supply tube that runs to a dishwasher adjacent to the sink, and the cold water riser may have a saddle valve and tubing running to a refrigerator icemaker.

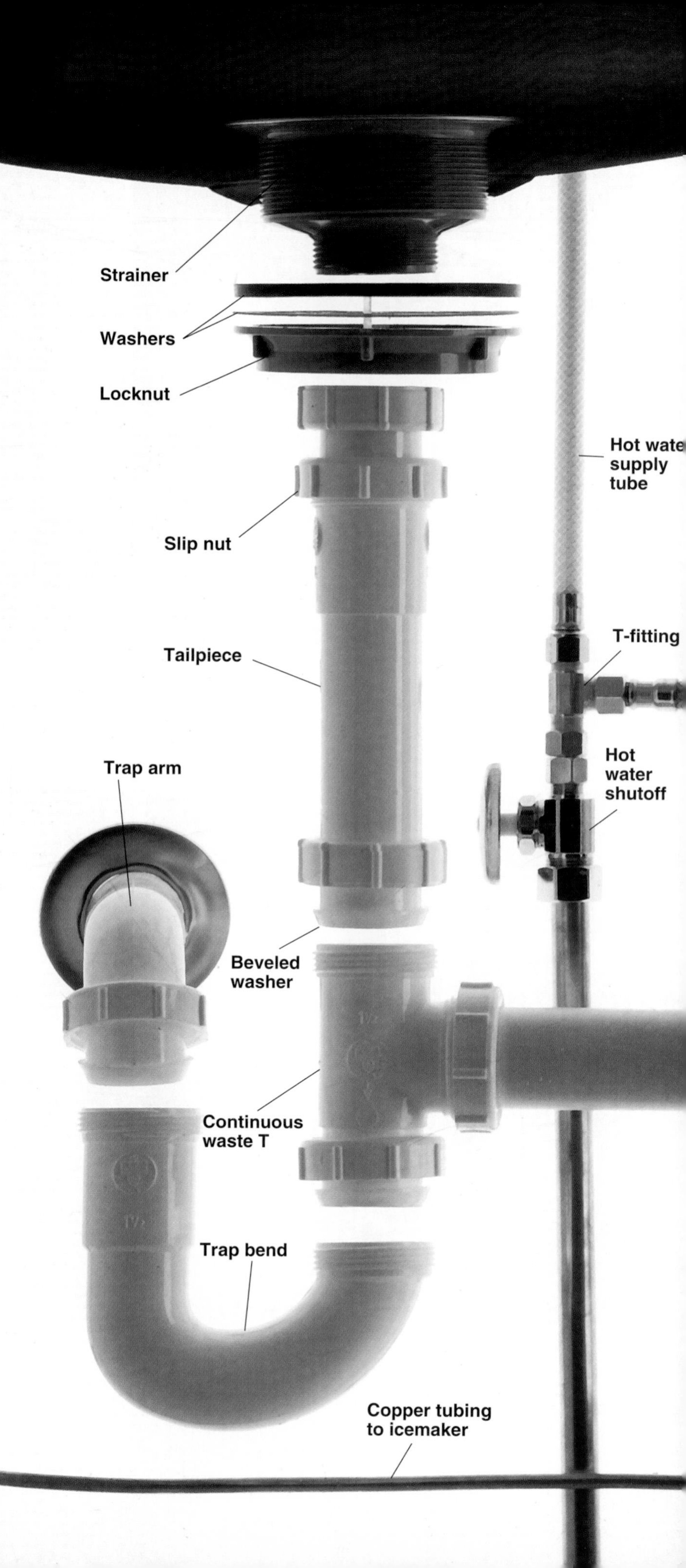

Dishwasher drain is looped up through an air gap device on the countertop (top), which prevents a plugged sink drain from backing up into the dishwasher. The supply tube running from the hot water supply pipe is connected to the water valve inside the access panel at the bottom of the dishwasher (bottom).

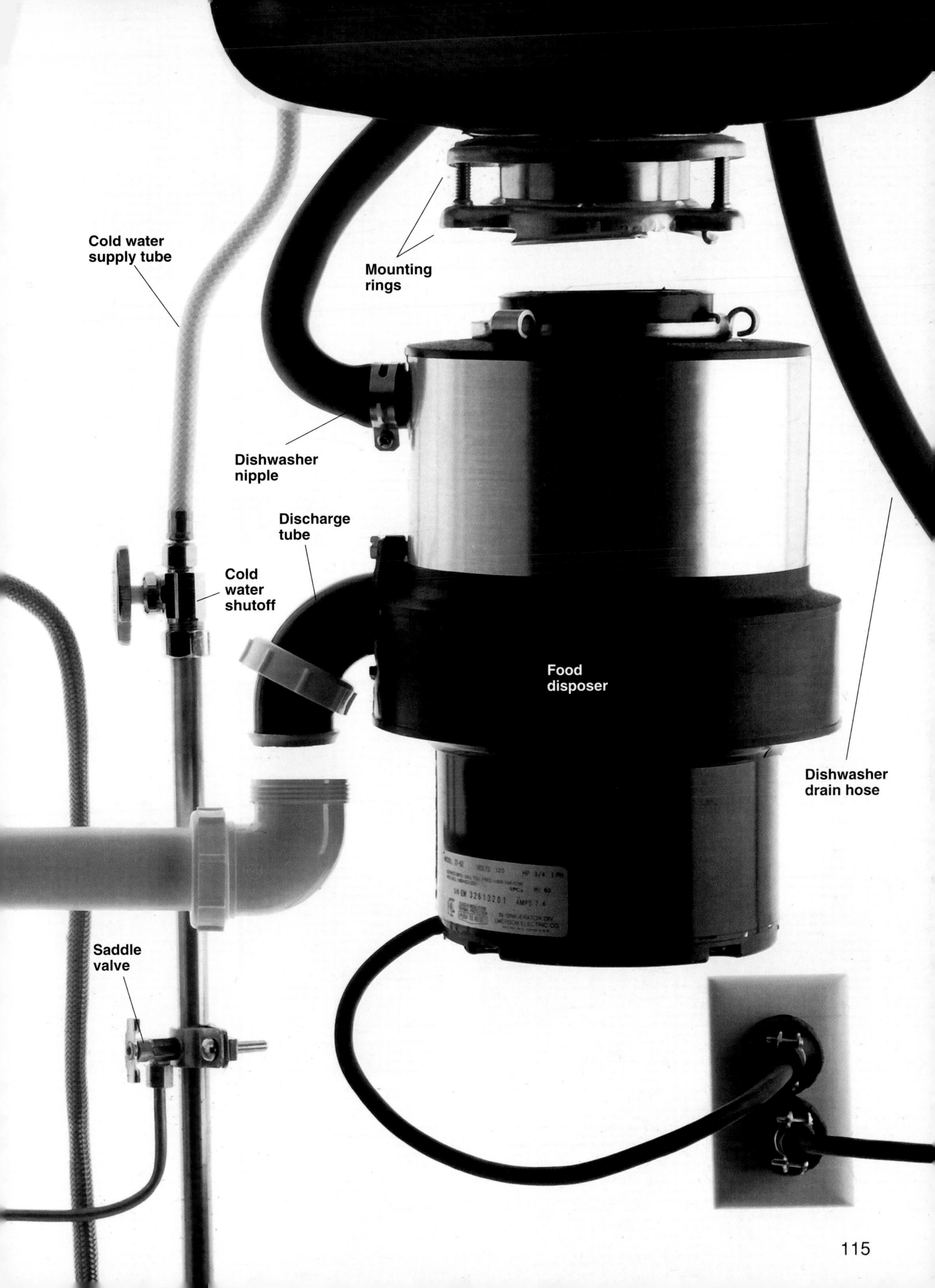
Cold water supply tube
Mounting rings
Dishwasher nipple
Discharge tube
Cold water shutoff
Food disposer
Dishwasher drain hose
Saddle valve

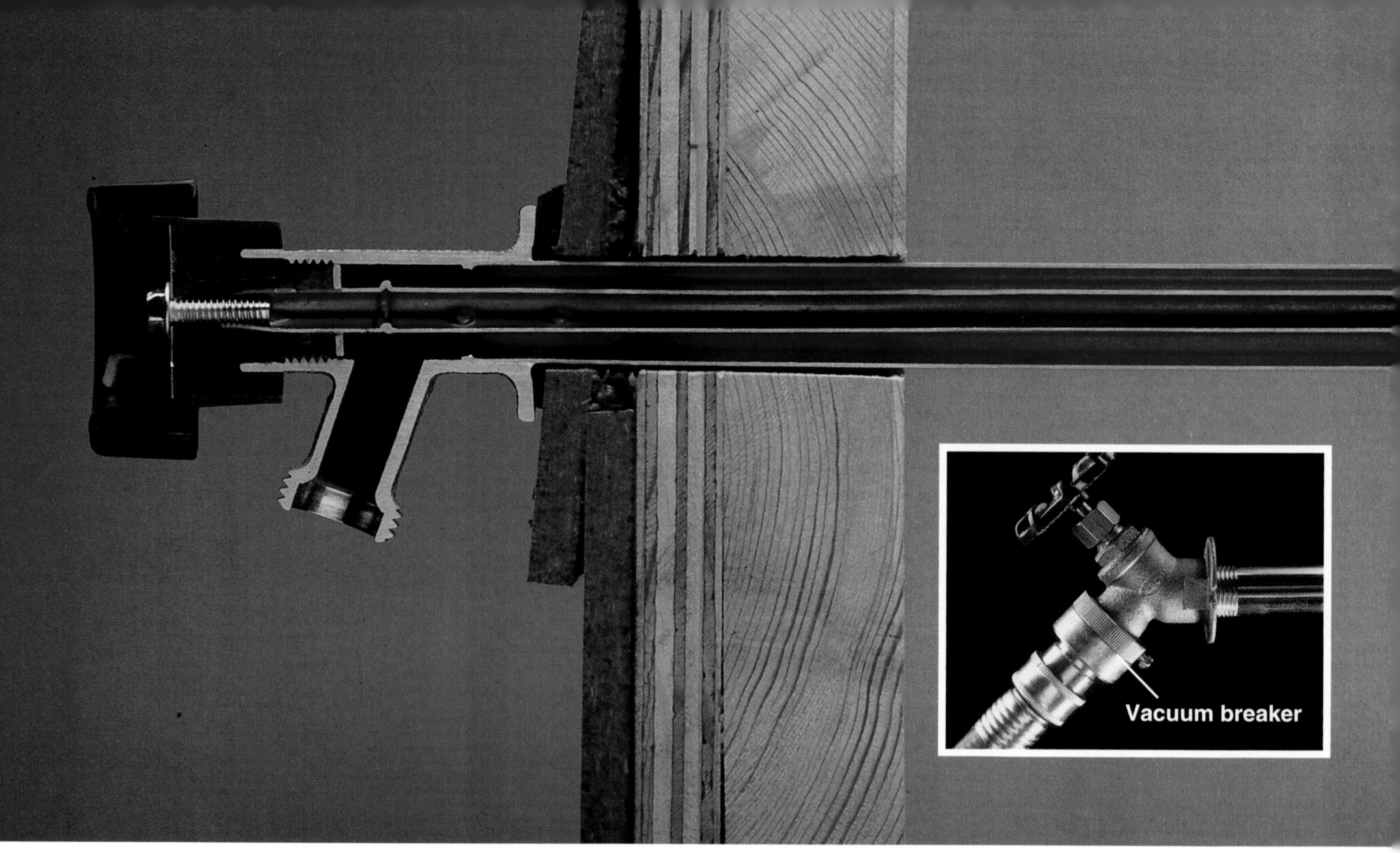

Sillcock is mounted against the header joist. Frost-free models have a long stem that reaches 6" to 30" inside the house to protect the valve from freezing temperatures. A sillcock should angle downward slightly toward the outdoors to provide drainage. The sillcock is connected to a nearby cold water supply pipe with a threaded

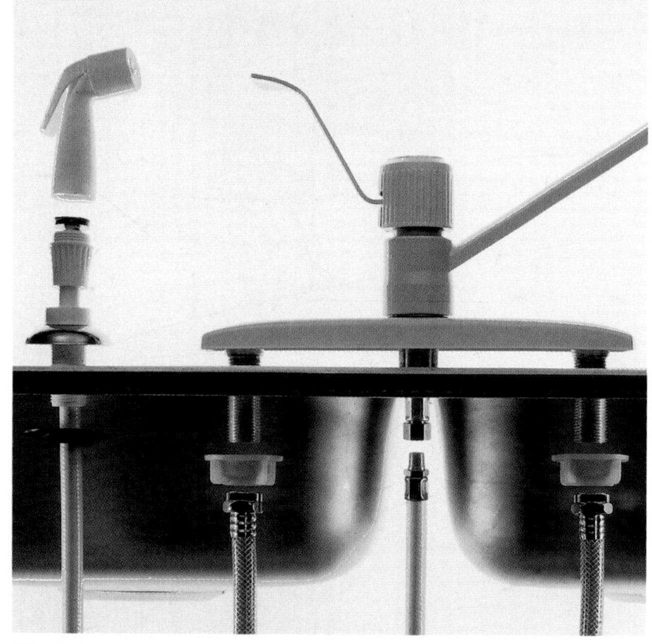

Sink faucet is connected to the water supply with supply tubes running from the tailpieces to shutoff valves on the water supply risers. On kitchen faucets with sprayers, an additional supply tube connects the mixing chamber of the faucet with a counter-mounted sprayer.

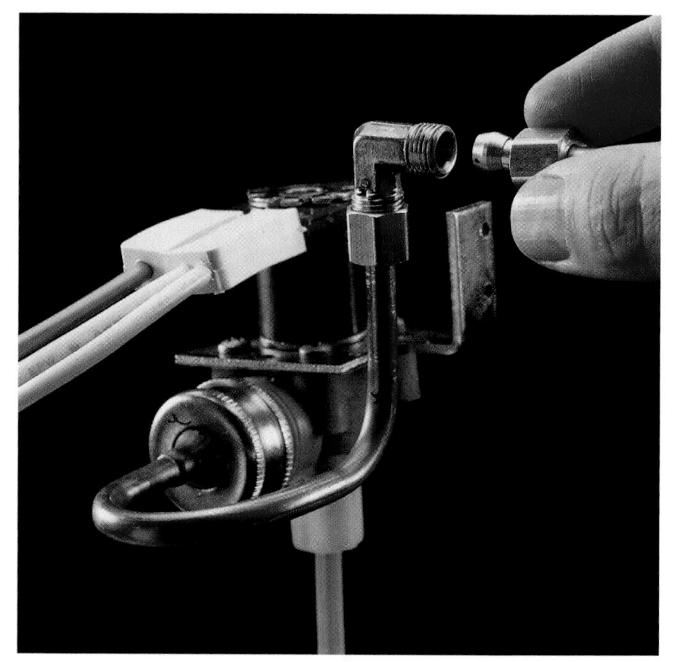

Refrigerator icemaker is connected with ¼" copper or plastic tubing running from a saddle valve on a cold water pipe (page 115) to the water valve tube on the refrigerator, using a ¼" compression elbow. Slide coupling nuts and compression rings over the tubes, and insert the tubes into the elbow. Tighten the nuts with channel-type pliers.

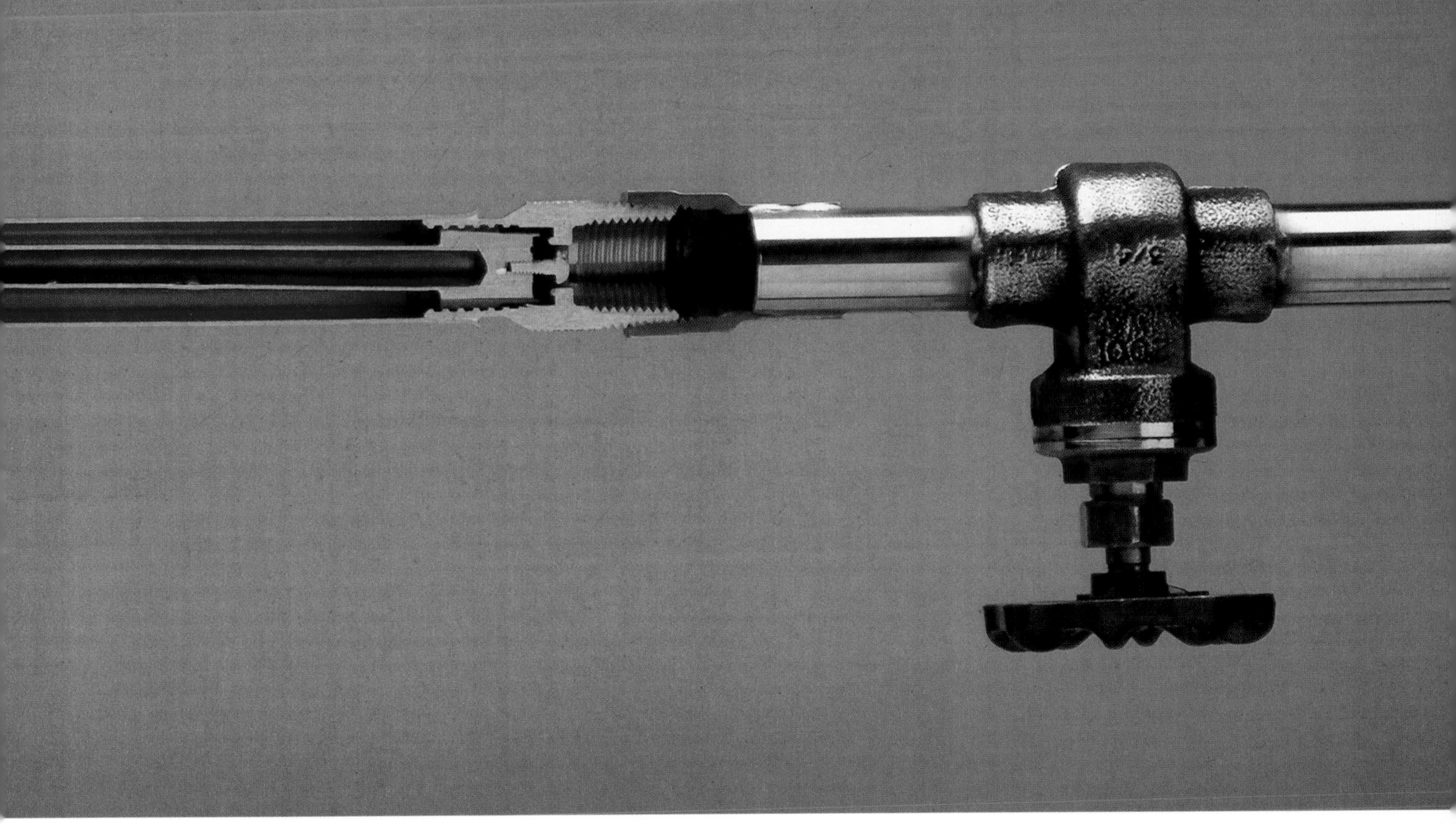

adapter, two lengths of soldered copper pipe, and a shutoff valve. A T-fitting (not shown) is used to tap into an existing cold water pipe. NOTE: If your sillcock does not have an anti-syphon feature, attach a threaded vacuum breaker to the hose bib (inset).

How to Connect a Water Heater

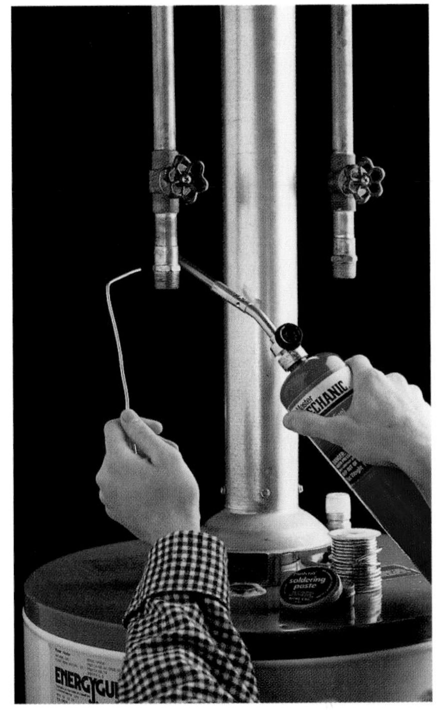

1 Solder full-bore valves and threaded male adapters to the cold and hot water pipes running to the water heater.

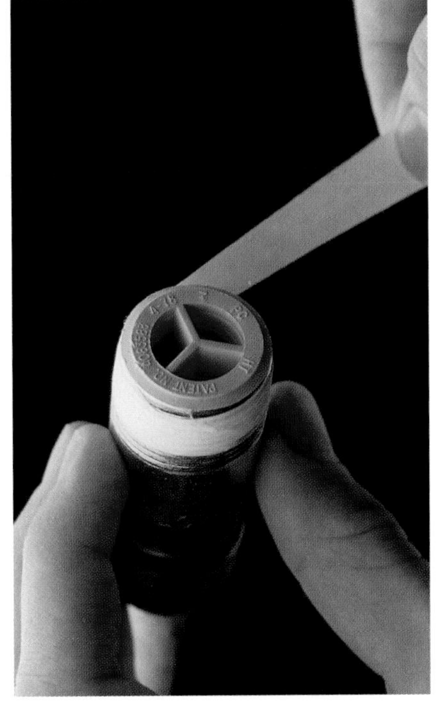

2 Wrap Teflon™ tape around the threads of two heat-saver nipples. The nipples are color-coded and have water-direction arrows to ensure proper installation.

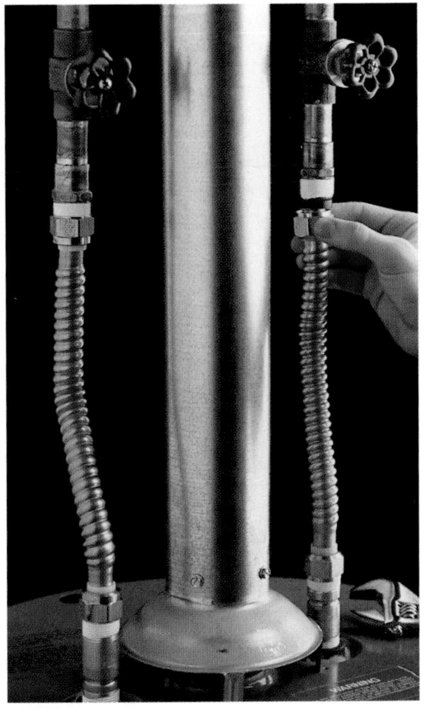

3 Connect the water lines to the heat-saver nipples with flexible water connectors. Tighten the fittings with an adjustable wrench.

INDEX

A

ABS pipes, 22, 28
  support intervals, 38
Access panel, 46

B

Backed-up floor drain, 89
Banded coupling, 32, 63, 98
  installing, 33, 69-70
Basement bathroom, plumbing, 58-63
Bathroom,
  basement bathroom, 58-63
  half bathroom, 64-65
  plumbing, 48-65
  stud thickness, 44
Bathtub,
  connecting drain, 113
  connecting faucets, 113
  illustration of system, 113
  installing DWV system, 54-55
  installing water supply lines, 57
  minimum clearance, 48
  minimum trap size, 38
  pipe size, 36
  water pressure, correct, 89
  whirlpool tub, 50, 54-55, 57
Booster pump, 37
Branch drains, replacing, 102-105
  see also: Drain pipes and drains
Branch pipes, 36
Bushings for pipes, 47

C

Capacity of water supply, determining, 88
Cast-iron pipes, 13
  connecting plastic pipes, 33
  support intervals, 38
  working with, 32-33
Ceiling, routing pipe through, 93
Chase, 90, 92, 96
Cleanout,
  fittings, 24, 38
  hub, 18
  replacing with new, 60-61
Clearances, minimum, 48, 64
Codes, *see:* Plumbing Codes
Concrete in basement bath, working with, 59, 63
Control valves, 45
Copper pipes, 12, 22
  connecting to galvanized pipes, 31
  support intervals, 38
  types, 22
  working with, 26-27
CPVC pipes, 22, 28
  support intervals, 38

D

Dielectric union, 31
Dishwasher,
  connecting, 114-115
  minimum trap size, 38
  pipe size, 36
Distribution pipes, sizes, 36
Drain cleanout, *see:* Cleanout
Drain pipes and drains, 35, 90-91
  connecting bathtub drain, 113
  connecting kitchen sink, 114-117
  connecting shower drain, 112
  connecting sink drain, 111
  connecting to main waste stack, 56
  cutting openings for, 51, 54
  fittings, 24-25
  identifying, 14
  installing, 51-55, 59-61, 65, 68-75
  mapping, 18-19
  pipe materials, 13, 22
  replacing old branch drains and vent pipes, 102-105
  replacing old toilet drain, 106-107
  sizes, 38
  slow drains, causes, 89
Drain trap, function, 13
Drain-waste-vent system, *see:* DWV system
Drop ear elbows, 65
Dry well, 78, 82
DWV (drain-waste-vent) system, 12, 13
  basement bath, installing, 58-63
  diagram, 24, 35
  fittings, 24-25, 47
  installing, 51-56
  mapping, 18-19
  pipes for, 13, 22, 28
  replacing old, 90-91
  testing, 40-41

E

Elbow fittings,
  drop ear, 65
  sweep 90°, 14, 24-25
Evaluating existing plumbing, 87-89

F

Faucet,
  connecting new, 111, 113-116
  measuring water pressure, 89
  see also: Sink
Filtration systems, in-home, 9
Fittings,
  identifying, 23-25
  to identify drain or vent pipes, 14
Flashing for waste-vent stack, 101
Flexible plastic pipe, working with, 30
Floor drain,
  backed-up, 89
  minimum trap size, 38
Food disposer, 114-115
Framing members,
  maximum hole or notch size, 46
  width for "wet" walls, 44, 50
Funny pipe, 83

G

Galvanized steel pipes, 12, 13
  connecting to copper pipes, 31
  replacing, 87, 108-109
  support intervals, 38
  working with, 31
Gray water, dry well for, 78
Groundwater supplies, 9, 10

H

Half bath, plumbing, 64-65
Hard water, effect on plumbing, 9
Holes in framing for pipe, 46
Home plumbing system,
  drain-waste-vent system, 12
  mapping, 14-19
  overview, 12-13
  water supply system, 12
Hose bibs, 16, 37
Hydrologic cycle, 9

I

In-home filtration systems, 9
In-home purification systems, 9
Iron pipes, *see:* Cast-iron pipes

K

Kitchen,
connecting sink system, 114-117
illustration of sink system, 114-115
plumbing, 66-77
stud thickness for, 44

L

Lagoon systems, 11
Landscape watering system, 79
components, 83
layout, 83
pipe materials, 22, 30
Laundry chute as pipe channel, 93
Layout for new plumbing, 44-47
Loop vent, installing, 66-67, 71-75

M

Main stack, locating, 14
Mapping home plumbing system, 14-19
Master bath, plumbing, 48-57
Metal protector plates, 46, 105
Microorganisms in water supply, 9
eliminating, 10
Mineral content of water, 9

N

Nitrates in water supply, 9
Notching framing for pipe, 46

O

Old plumbing, replacing, 86-117
evaluating existing system, 87-89
Outdoor water supply,
dry well, 78, 82
installing plumbing, 78-83
pipe materials, 22

P

PE pipes, 22, 78
working with, 30
Permit, obtaining, 35
Pipe cutter, using, 26
Pipes,
common materials, 12, 13, 22-23
mapping, 14-19
offset, locating, 17
planning pipe routes, 90, 92-95
sizes, 36, 38
testing, 40-41
working with, 26-33
Plastic pipes, 12, 13
connecting to cast-iron pipes, 33
types of plastic, 22-23
working with flexible, 30
working with rigid, 28-29
Plumbing codes,
permits, 35
understanding, 34
zones, 34
Plumbing symbols, 15
Pollutants in water supply, 9
Pressure-reducing valve, 37
Pressure test, 40-41
installing T-fittings for, 47
Primer for plastic pipes, 28
Protector plates, 46, 105
P-trap, 111
Purification systems, in-home, 9
PVC pipes, 22, 28
support intervals, 38

R

Rattling pipes, preventing, 47
Recycling waste water, 8, 9, 11
Reducing Y-fitting, 61
Refrigerator icemaker, 114, 116
Replacing old plumbing, 86-117
Reservoirs, man-made, 10
Rigid plastic pipes,
types, 22
working with, 28-29
Roof, installing new waste-vent stack, 100
flashing, 101
Rust,
stains on fixtures, causes, 89
on tools, preventing, 21

S

Septic systems, 9, 11
Sewage treatment plants, 9, 11
Shower,
connecting new, 112
illustration of system, 112
in basement bathroom, 58-63
installing DWV system, 54-55
minimum size, 48
minimum trap size, 38
pipe size, 36
Shutoff valves, 76-77
Sidewalk, running pipe under, 81
Sillcock, 16
connecting, 116-117
pipe size, 36
shutoff valve, 36
Sink,
connecting new, 111, 114, 116
dry well for, 78, 82
illustration of system, 114-115
in basement bathroom, 58-63
in garage, 80-82
installing DWV system, 51-53, 65, 67-75
installing near existing lines, 64-65
installing water supply lines, 57
island sink, 66-67, 71-75
minimum clearance, 48
minimum trap size, 38
pipe size, 36
water pressure, correct, 89
Site plan, 35
Slow drains, causes, 89
Snap cutter, 21, 32, 97
Soldering copper pipes, 27
Sprinkler system, *see:* Landscape watering system
Steel pipes, *see:* Galvanized steel pipes
Stud thickness for "wet" walls, 44, 50
Supply lines, *see:* Water supply system
Supply riser diagram, 35
Supply tubes, 36
Support materials for pipes, 23, 47
intervals, 38, 56, 65, 108
Surface water supplies, 9, 10
Sweep 90° elbow fitting, 14, 24-25
Symbols for mapping plumbing, 15

T

Testing new pipes, 40-41
T-fittings, 14, 24-25, 30
for pressure test, 47
Toilet,
connecting new, 110
illustration of system, 110
in basement bathroom, 58-63
installing DWV system, 51-53, 65
installing near existing lines, 64-65
minimum clearance, 48
pipe size, 36, 38
replacing old drain, 106-107
venting, 39
Trap size,
minimum, 38
trap-to-vent distances, 39

U

Underground water supply,
pipe materials, 22, 30
vacuum breaker for, 37

V

Vacuum breakers, 37
Valves,
control valves, 45
locating and mapping, 15
required valves, 36
types, 23
Vent pipes, 35
and slow drains, 89
connecting to main waste stack, 56, 69-70, 105
fittings, 24-25
function, 13
identifying, 14
installing, 53-55, 59-60, 62-63, 65, 69-75
loop vent, 66-67, 71-75
mapping, 18-19
pipe materials, 13, 22
replacing old, 90-91, 102-105
sizes, 39
trap-to-vent distances, 39
wet venting, 39, 58
see also: Waste-vent stack

W

Washing machine,
minimum trap size, 38
Waste-vent stack,
auxiliary, 18
connecting to new DWV pipes, 56, 69-70, 105
mapping, 18-19
pipe materials, 13, 22
replacing cleanout, 60-61
replacing old, 90, 96-101
Waste water, recycling, 9, 11
Water cycle, understanding, 8-9
Water hammer arrester, 37
Water heater, 12, 16, 36
connecting, 117
Water meter, 12, 16, 36
Water pressure,
and water tower height, 10
measuring, 89
modifying, 37
Water quality, determining, 9
Water softener, 9
Water supply system, 12
branch pipes, 36
connecting bathtub faucets, 113
connecting kitchen sink, 114-116
connecting shower pipes, 112
connecting sink faucets, 111
connecting whirlpool faucets, 113
determining capacity, 88
fittings, 23
installing new supply pipes, 57, 63, 65, 76-77
installing outdoor system, 80-83
locating existing lines, 14
mapping, 16-17, 35
pipe materials, 12, 22, 26
replacing old, 90-91, 108-109
sizing of pipes, 36
supply tubes, 36
testing, 40-41
valves, 23
Water towers, 10
Water treatment plants, 9, 10
Wells, 8, 9
Wet venting, 39, 58
"Wet" walls, 94
framing, 44, 50
Whirlpool tub,
connecting faucets, 113
installing DWV system, 54-55
installing water supply lines, 57
strengthening floor, 50, 54
Winterizing outdoor plumbing, 81-82

Y

Y-fittings, 14, 24-25
reducing, 61

Z

Zones for plumbing codes, 34